Transition Metal-Catalyzed Benzofuran Synthesis

Transition Metal-Catalyzed Benzofuran Synthesis

Transition Metal-Catalyzed Heterocycle Synthesis Series

Xiao-Feng Wu and Yahui Li

Leibniz-Institut für Katalyse e.V.
an der Universität Rostock (LIKAT),
Rostock, Germany

ELSEVIER   elsevier.com

Elsevier
Radarweg 29, PO Box 211, 1000 AE Amsterdam, Netherlands
The Boulevard, Langford Lane, Kidlington, Oxford OX5 1GB, United Kingdom
50 Hampshire Street, 5th Floor, Cambridge, MA 02139, United States

Notices
Knowledge and best practice in this field are constantly changing. As new research and
experience broaden our understanding, changes in research methods, professional practices,
or medical treatment may become necessary.

Practitioners and researchers must always rely on their own experience and knowledge in
evaluating and using any information, methods, compounds, or experiments described herein.
In using such information or methods they should be mindful of their own safety and the safety
of others, including parties for whom they have a professional responsibility.

To the fullest extent of the law, neither the Publisher nor the authors, contributors, or editors,
assume any liability for any injury and/or damage to persons or property as a matter of products
liability, negligence or otherwise, or from any use or operation of any methods, products,
instructions, or ideas contained in the material herein.

British Library Cataloguing-in-Publication Data
A catalogue record for this book is available from the British Library

Library of Congress Cataloging-in-Publication Data
A catalog record for this book is available from the Library of Congress

ISBN: 978-0-12-809377-1

For Information on all Elsevier publications
visit our website at https://www.elsevier.com/books-and-journals

www.elsevier.com • www.bookaid.org

Publisher: John Fedor
Acquisition Editor: Emily McCloskey
Editorial Project Manager: Jill Cetel
Production Project Manager: Vijayaraj Purushothaman

Typeset by MPS Limited, Chennai, India

CONTENTS

CONTENTS

CHAPTER 1

Introduction

Benzofurans is a class of heterocyclic compound consisting of fused benzene and furan rings. As naturally occurring compounds, benzofurans have been used in pharmaceuticals, etc. as well (Scheme 1.1). Due to their potential applications, numerous methodologies have been developed during the past few decades for their selective and effective preparation [1].

In this volume, the progress of transition metal-catalyzed furan synthesis is discussed. Based on the catalysts applied, this text is organized by Pd, Cu, Ru, Rh, Pt, etc., i.e., the catalyzed reactions for the synthesis of benzofuran. Some traditional procedures have been included and discussed here as well to ensure readability.

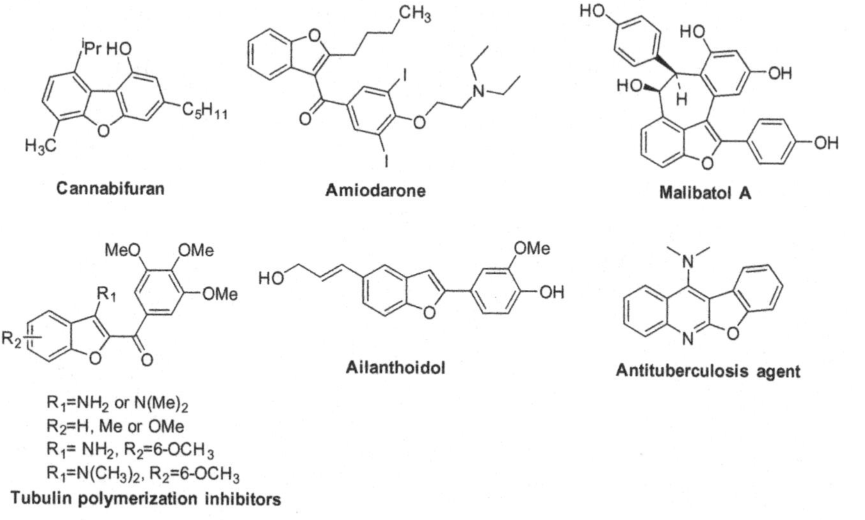

Cannabifuran

Amiodarone

Malibatol A

R_1=NH$_2$ or N(Me)$_2$
R_2=H, Me or OMe
R_1= NH$_2$, R_2=6-OCH$_3$
R_1=N(CH$_3$)$_2$, R_2=6-OCH$_3$
Tubulin polymerization inhibitors

Ailanthoidol

Antituberculosis agent

Scheme 1.1 Selected examples of benzofuran derivatives.

Transition Metal-Catalyzed Benzofuran Synthesis. DOI: http://dx.doi.org/10.1016/B978-0-12-809377-1.00001-2

REFERENCES

[1] a) Hayta, S. A.; Arisoy, M.; Arpaci, O. T.; Yildiz, I.; Aki, E.; Ozkan, S., et al. *Eur. J. Med. Chem.* **2008**, *43*, 2568–2578.

b) Yeung, K.-S. *Heterocycl. Chem.* **2012**, *29*, 47–76.

c) Kamal, M.; Shakya, A. K.; Jawaid, T. *Int. J. Med. Pharm. Sci.* **2011**, *1*, 1–15.

d) Verma, A.; Pandeya, S. N.; Sinha, S. *Int. J. Res. Ayurveda Pharm.* **2011**, *2*, 1110–1116.

e) Deshpande, R.; Bhagawan Raju, M.; Parameshwar, S.; Shanth Kumar, S. M.; Appalaraju, S.; Yelagatti, M. S. *Am. J. Chem.* **2012**, 115–120.

f) Lin, S.-Y.; Chen, C.-L.; Lee, Y.-J. *J. Org. Chem.* **2003**, *68*, 2968–2971.

g) Zhu, R.; Wei, J.; Shi, Z. *Chem. Sci.* **2013**, *4*, 3706–3711.

h) Li, B.; Yue, Z.; Xiang, H.; Lv, L.; Song, S.; Miao, Z., et al. *RSC Adv.* **2014**, *4*, 358–364.

CHAPTER 2

Pd-Catalyzed Benzofuran Synthesis

In 1986, Marinelli and coworkers developed a palladium-catalyzed reaction of *o*-iodophenols with 1-alkynes. This method represents a very useful procedure for the synthesis of substituted-benzo[*b*]furan (Scheme 2.1) [1]. In this procedure, good yields of the benzo[*b*]furan can be obtained in the presence of Pd(OAc)$_2$(PPh$_3$)$_2$ (2 mol%), CuI (4 mol%), and piperidine at room temperature, or 60°C. This method also has two main advantages. One is mild reaction conditions, and the other is compatibility with a variety of functional groups, both in the phenolic and the alkyne moiety.

Later, Kundu and coworkers reported palladium-catalyzed hetero-annulation of acetylenic compounds for synthesis of benzo[*b*]furan (Scheme 2.2) [2]. They synthesized a number of 2-substituted benzofurans from readily accessible starting materials under relatively mild conditions and with fair to excellent yields. Some of the synthesized compounds were easily converted to known natural products or compounds of biological interest.

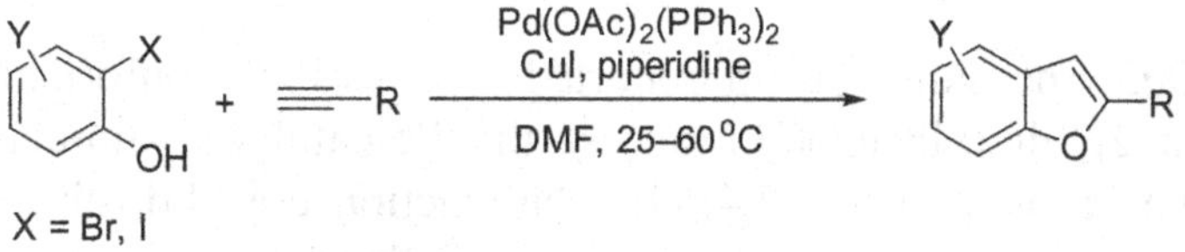

Scheme 2.1 *Palladium-catalyzed synthesis of benzo[b]furan from alkynes.*

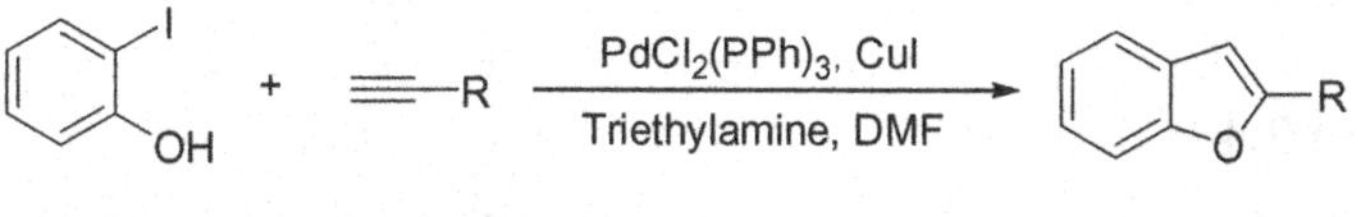

Scheme 2.2 *Palladium-catalyzed synthesis of benzo[b]furan from alkynes.*

Transition Metal-Catalyzed Benzofuran Synthesis. DOI: http://dx.doi.org/10.1016/B978-0-12-809377-1.00002-4

Scheme 2.2

Synthesis of benzofuran-2-ylmethanol. A mixture of o-iodophenol (2 mmol), PdCl$_2$(PPh$_3$)$_2$ (0.07 mmol), CuI (0.26 mmol), and triethylamine (4 mmol) was stirred in dimethylformamide (5 mL) under nitrogen for 1 h. Prop-2-ynyl alcohol (4 mmol) was added, and the mixture was stirred at room temperature for 1 h more and heated at 60°C for 16 h. The mixture was then cooled, poured into water (100 mL), and extracted with dichloromethane (3×50 mL). The combined organic extracts were washed with 5 mol dm^{-3} sodium hydroxide (3×100 mL) and water, dried (MgSO$_4$), and purified by chromatography on neutral alumina.

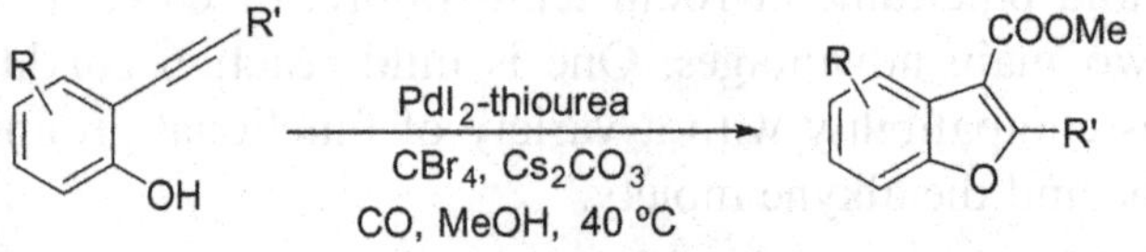

Scheme 2.3 Palladium − thiourea catalyzed carbonylative annulation of o-hydroxylarylacetylenes.

Yang and coworkers developed an effective cocatalysis system (PdI$_2$-thiourea and CBr$_4$) for carbonylative cyclization of both electron-rich and electron-deficient o-hydroxylarylacetylenes to the corresponding methyl benzo[b]furan-3-carboxylates (Scheme 2.3) [3]. The application of this cocatalyst system to a silyl linker-based solid-phase benzo[b]furan-3-carboxylate synthesis proved to be as effective as in solution phase.

In 2003, the same group reported synthesis of conformationally restricted 2,3-diarylbenzo[b]furan by the Pd-catalyzed annulation of o-alkynylphenols (Scheme 2.4) [4]. The reaction used Pd$_2$(dba)$_3$ as the catalyst, bipyridine as the ligand, and gave the desired results in good yield. The aryl iodides with electron-withdrawing groups gave the best results, presumably due to their favorable effect on the oxidative addition to the Pd^0 complex.

Scheme 2.4

General Procedure for the Palladium-Catalyzed Formation of 2,3-Biarylbenzo[b]furans. Pd$_2$(dba)$_3$ (46 mg, 0.05 mmol, 5 mol%) was added to a solution of acetonitrile (3.0 mL), ArI (2.0 mmol, 200 mol%), bpy (16 mg, 0.1 mmol, 10 mol%), and K_2CO_3 (552 mg, 4.0 mmol) and the mixture was stirred at 50°C for 1 h. To this solution a solution of o-alkylnylphenol (1.0 mmol, 100 mol%) in acetonitrile (2 mL) was added, and the reaction mixture was stirred at 50°C for 5 h under Ar_2 atmosphere. The reaction mixture was then concentrated, and the residue was filtered through a silica gel pad and eluted with EtOAc. The filtrate was concentrated, and the residue was purified by flash chromatography on silica gel to give the corresponding cyclization product.

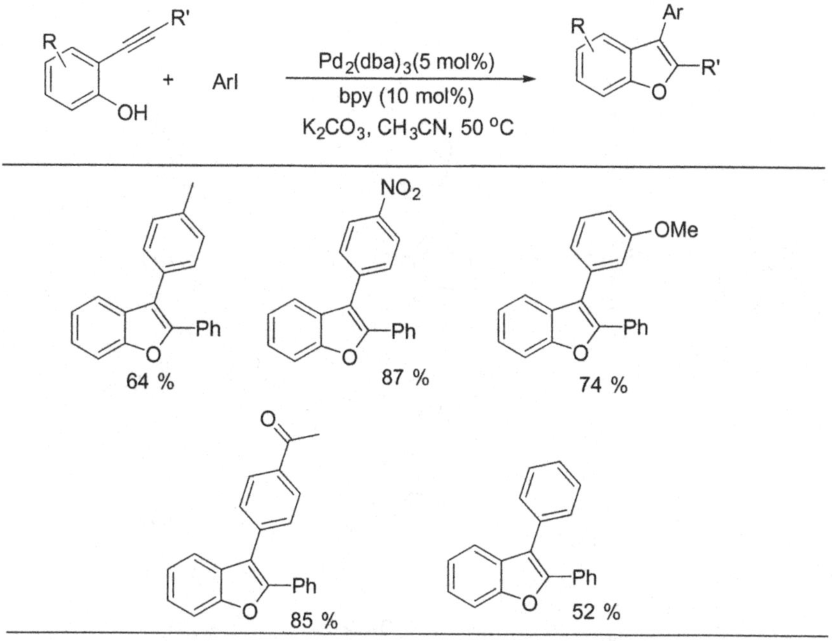

Scheme 2.4 Synthesis of 2,3-diarylbenzo[b]furan by the Pd-catalyzed.

The key step of the previous three approaches is attack of a nucleophilic phenol oxygen atom onto an activated C−C multiple bond. In 2004, Willis et al. reported alternative palladium-catalyzed cyclization in which the nucleophilic oxygen atom of an enolate was coupled with a halo-substituted arene ring (Scheme 2.5) [5]. In the presence of $Pd_2(dba)_3$ and the ligand DPEphos, the corresponding benzofurans were obtained by intramolecular C − O bond formation between 1-(2-haloaryl)ketones and aryl halides. Both cyclic and acyclic ketones are efficient substrates.

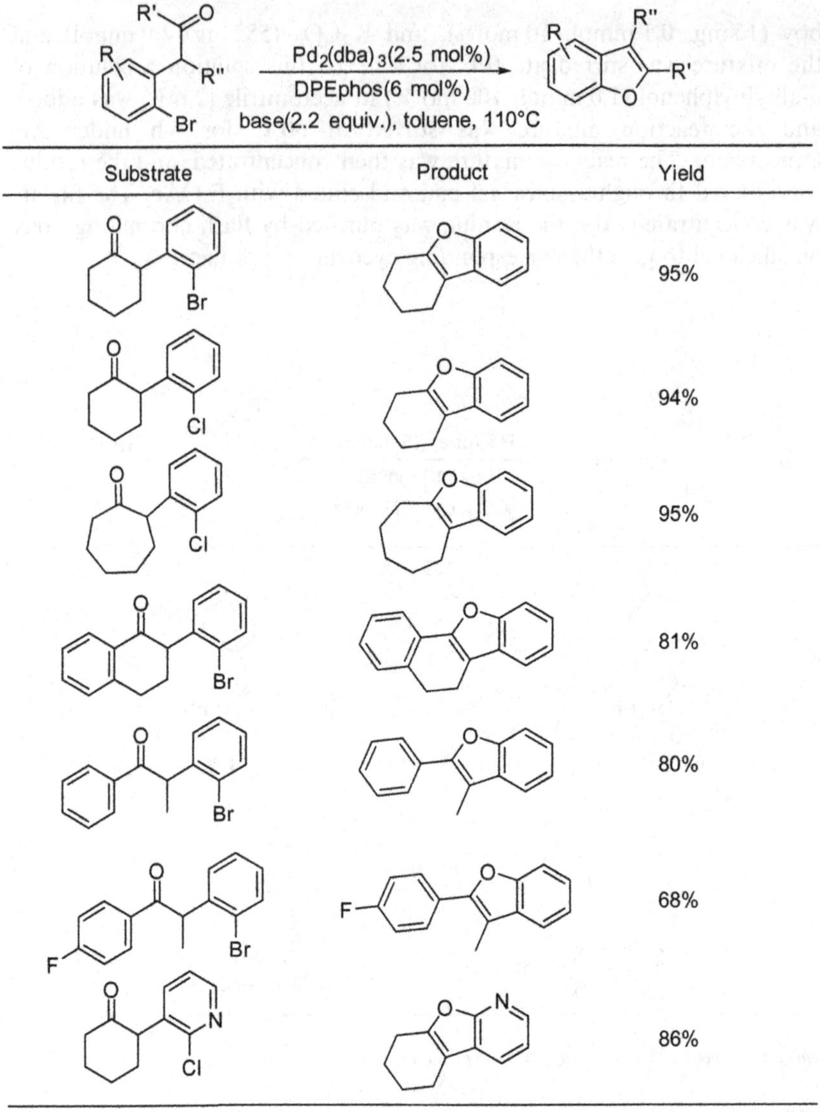

Pd2(dba)3(2.5 mol%)
DPEphos(6 mol%)
base(2.2 equiv.), toluene, 110°C
Substrate
Product
Yield
95%
94%
95%
81%
80%
68%
86%

Scheme 2.5 Palladium-catalyzed intramolecular o*-arylation of enolates.*

Scheme 2.5

Synthesis of 1,2,3,4-tetrahydro-dibenzofuran. Cesium carbonate (0.18 g, 0.56 mmol) was added to a flask charged with $Pd_2(dba)_3$ (9 mg, 0.01 mmol) and DPEphos (13 mg, 0.02 mmol) under nitrogen. The reagents were suspended in anhydrous toluene (1 mL), 2-(2-bromophenyl) cyclohexanone (0.10 g, 0.40 mmol) was added, and the reaction was heated to 100°C for 20 h. After cooling the reaction mixture was filtered through a plug of celite and the filtrate reduced in vacuo. The residue was purified *via* flash chromatography (petroleum ether) to yield the title compound (64 mg, 95%) as a colorless oil.

In 2004, Pan and coworkers reported the synthesis of benzofurans in ionic liquid by a PdCl$_2$-catalyzed intramolecular Heck reaction (Scheme 2.6) [6]. The reaction was carried out in ionic liquid

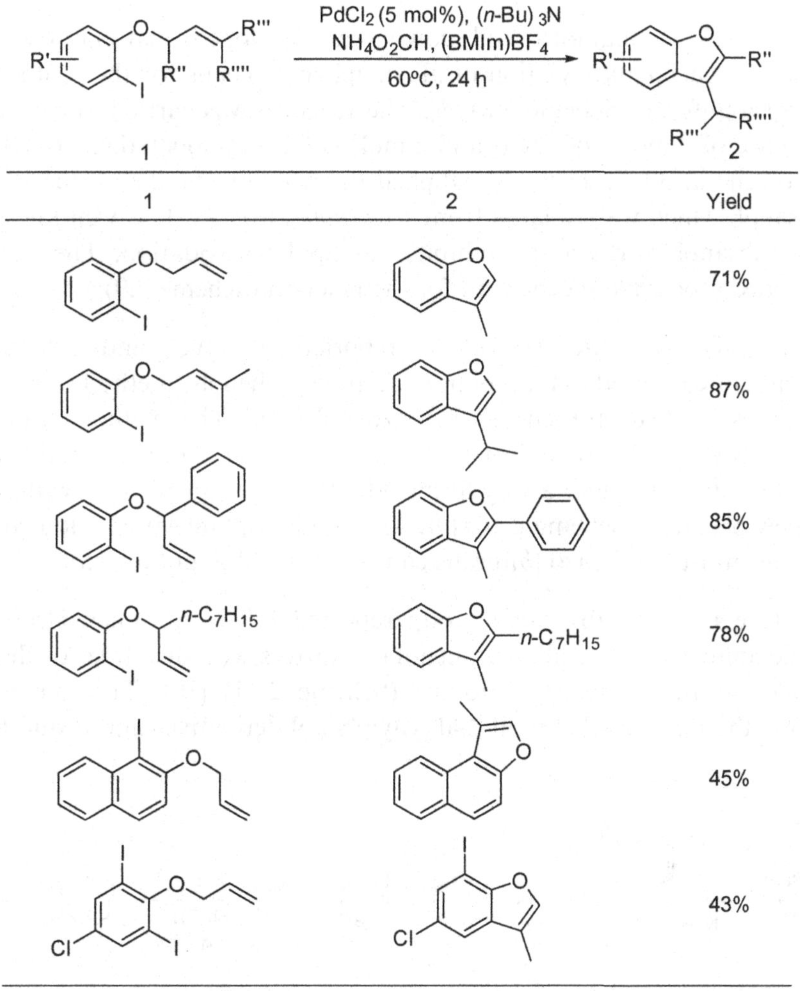

1	2	Yield
		71%
		87%
		85%
		78%
		45%
		43%

Scheme 2.6 Synthesis of benzo[b]furans in ionic liquid by a PdCl$_2$-catalyzed intramolecular Heck reaction.

(1-*n*-butyl-3-methylimidazolium tetraborate) and substituted-benzo[*b*] furans were obtained in modest to satisfactory yields. Interestingly, the ionic liquid containing Pd catalyst can be used four times with little loss of activity.

In 2002, Li and coworkers developed palladium-mediated reactions for the synthesis of all four possible benzo[4,5]-furopyridine tricyclic heterocycles, benzo[4,5]furo[2,3-*b*]pyridine, benzo[4,5]furo[2,3-*c*]pyridine, benzo[4,5]furo[3,2-*c*]pyridine, and benzo[4,5]-furo[3,2-*b*]pyridine, from the simple substances 2-chloro-3-iodopyridine, 3-chloro-4-stannyl-pyridine, 4-chloro-3-iodopyridine, and 2-chloro-3-hydroxypyridine, respectively (Scheme 2.7) [7].

In 2005, Yamamoto and coworkers developed platinum-olefin-catalyzed carbo-alkoxylation of alkynylphenyl acetals for the synthesis of benzo[*b*]furan (Scheme 2.8) [8]. The reaction was carried out in the presence of 2 mol% of PtCl$_2$ and 8 mol% of 1,5-cyclooctadiene (COD) in toluene at 30°C. They also applied the reaction for the synthesis of vibsanol, which was isolated from *Viburnum awabuki*. It is well known that vibsanol works as an inhibitor of lipid peroxidation. They also provided a possible mechanism for the reaction (Scheme 2.9).

In 2006, Li and coworkers reported a novel and selective palladium-catalyzed annulation of 2-alkynylphenols method for the synthesis of 2-substituted 3-halobenzo [*b*]furans. The method affords products with a halide (Cl or Br) at the 3 position. Adding a halide at the 3 position provides a way to introduce new groups for the synthesis of new products (Scheme 2.10) [9]. Also, under optimized reaction conditions, most halobenzo [*b*]furans can be obtained in good yield.

More recently, the same group reported PdCl$_2$-promoted electrophilic annulation of 2-alkynylphenol derivatives with disulfides or diselenides in the presence of iodine (Scheme 2.11) [10]. This method allows the reactions between 2-alkynylphenol derivatives and disulfides

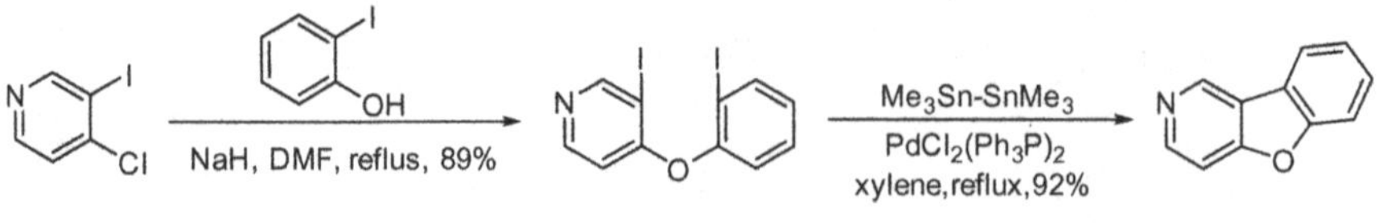

Scheme 2.7 Synthesis of benzo[4,5]furopyridines via palladium-mediated reactions.

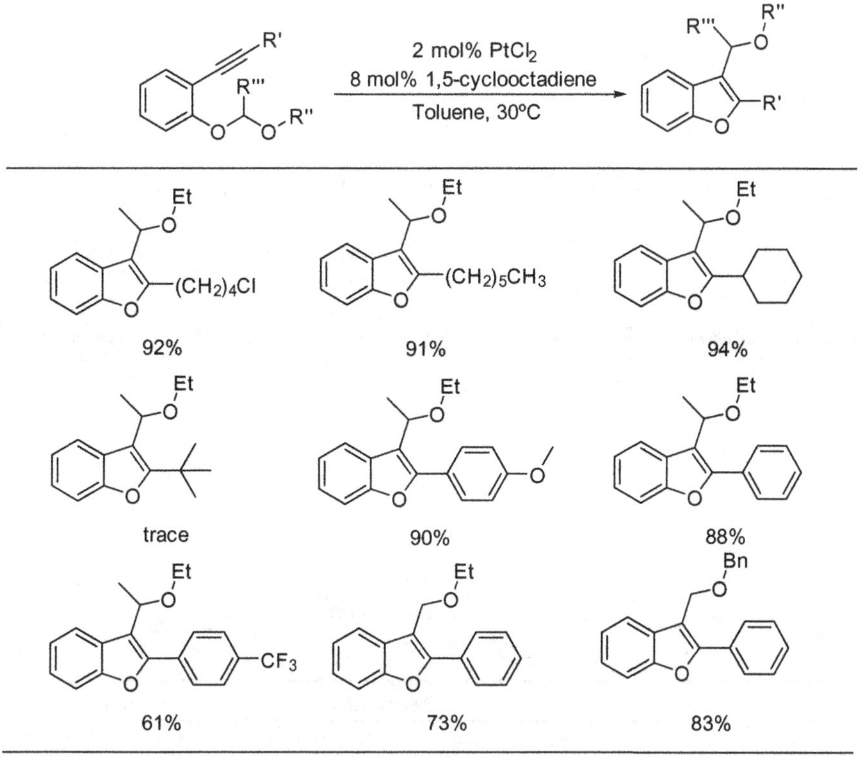

Scheme 2.8 Synthesis of 2,3-disubstituted benzofurans by platinum-olefin-catalyzed arboalkoxylation of alkynyl-phenyl acetals.

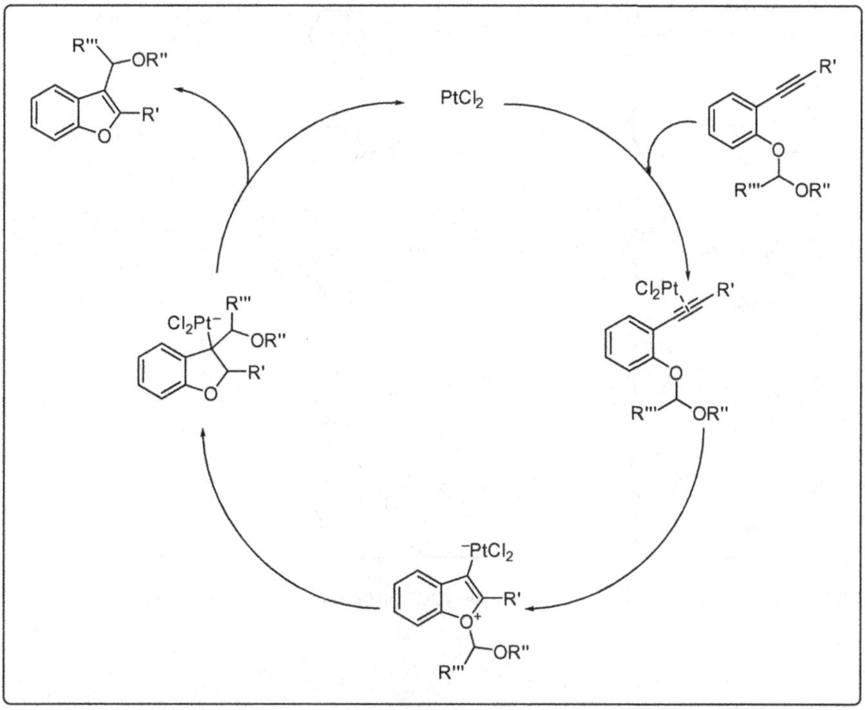

Scheme 2.9 Possible mechanism.

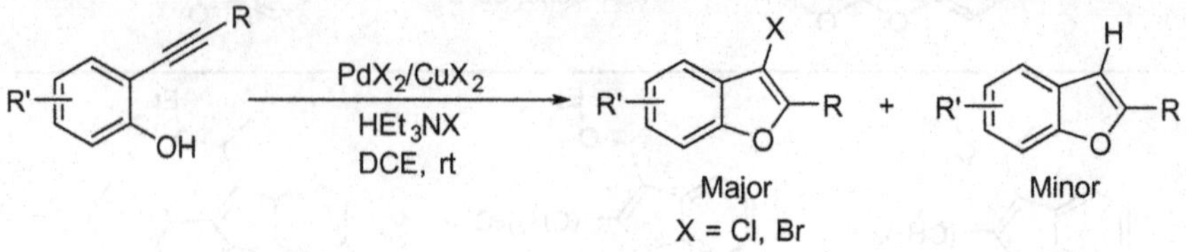

Scheme 2.10 Palladium-catalyzed annulations of 2-alkynylphenols to form 2-substituted 3-halobenzo[b]furans.

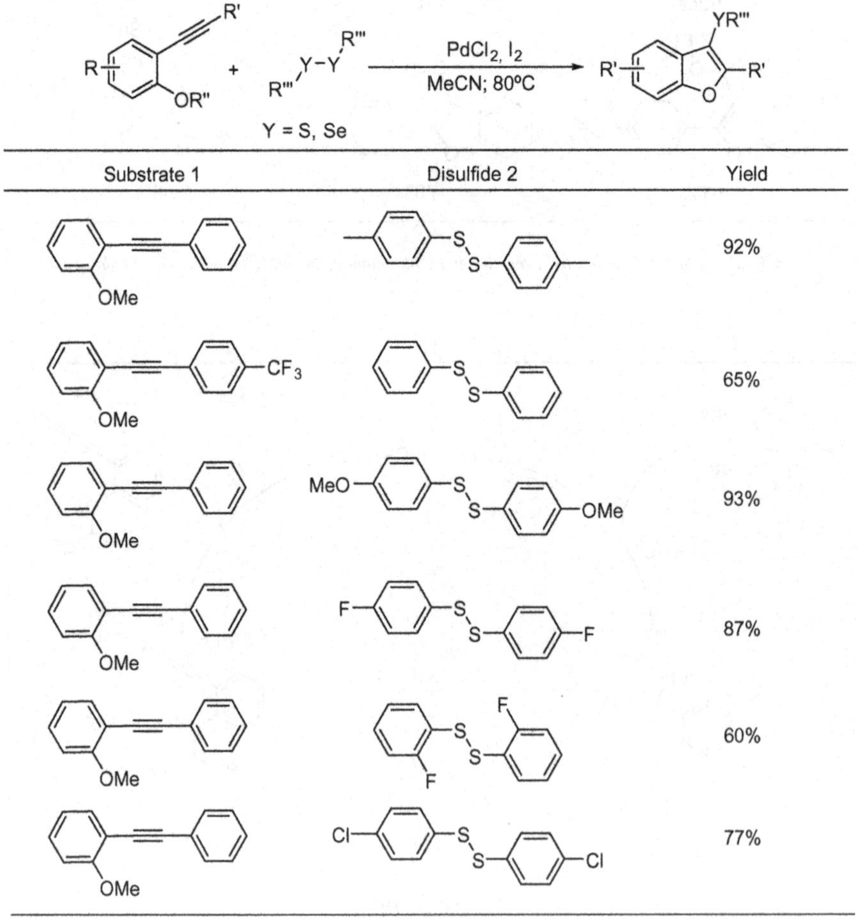

Substrate 1	Disulfide 2	Yield
		92%
		65%
		93%
		87%
		60%
		77%

Scheme 2.11 PdCl₂-catalyzed electrophilic annulation of 2-alkynylphenol derivatives with disulfides or diselenides.

(or diselenides) for the synthesis of 3-chalcogen-benzo[*b*]furans. The key step of the reaction is generation of RYI (Y = S, Se) in situ from the reaction of RYYR with I_2. $PdCl_2$ can improve the reaction. In the presence of $PdCl_2$ and I_2, a variety of 2-alkynylphenol derivatives underwent annulation reactions with disulfides or diselenides to successfully result in corresponding 3-sulfenylbenzofurans and 3-selenenylbenzofurans in moderate to excellent yields.

Scheme 2.11

A mixture of 2-alkynylphenol derivative 1 (0.2 mmol), disulfide or diselenide 2 (0.1 mmol), I_2 (2 equiv.), and $PdCl_2$ (10 mol%) in MeCN (2 mL) was stirred at 80°C for the indicated time until complete consumption of starting material as monitored by TLC and GC–MS analysis. After the reaction was finished, the mixture was poured into ethyl acetate, which was washed with saturated NaS_2O_3 and extracted with diethyl ether. The organic layers were dried over anhydrous Na_2SO_4 and evaporated under vacuum. The residue was purified by flash column chromatography (hexane/ethyl acetate) to create the products.

In 2008, Aurrecoechea and coworkers developed a new Pd-catalyzed tandem intramolecular oxypalladation/Heck-type coupling between 2-alkynylphenols and alkenes, leading to 3-(1-alkenyl) benzofurans (Scheme 2.12) [11]. In the presence of $PdCl_2$ and KI, 2-alkynylphenols reacted with alkenes, resulting in the desired benzofuran with excellent to good yields. The reaction has some features including (1) a high degree of structural diversity, (2) effective participation of ketone derivatives without unwanted hydroarylation-type side reactions, and (3) tolerance of substitution at the electron-deficient olefin β–position.

Scheme 2.12

To a solution of 2-alkynylphenols (0.24 mmol) in DMF (2 mL) was added $PdCl_2$ (0.003 g, 0.012 mmol), KI (0.020 g, 0.120 mmol) and alkenes (1.44 mmol) and the mixture was stirred at 80°C for 20 h. The mixture was allowed to cool to 25°C and water was added. The mixture was extracted with EtOAc (three times), the combined organic layers were dried (Na_2SO_4), and the solvent was removed. The residue was purified by chromatography (silica gel, 95:5 hexane/EtOAc) to afford the desired product.

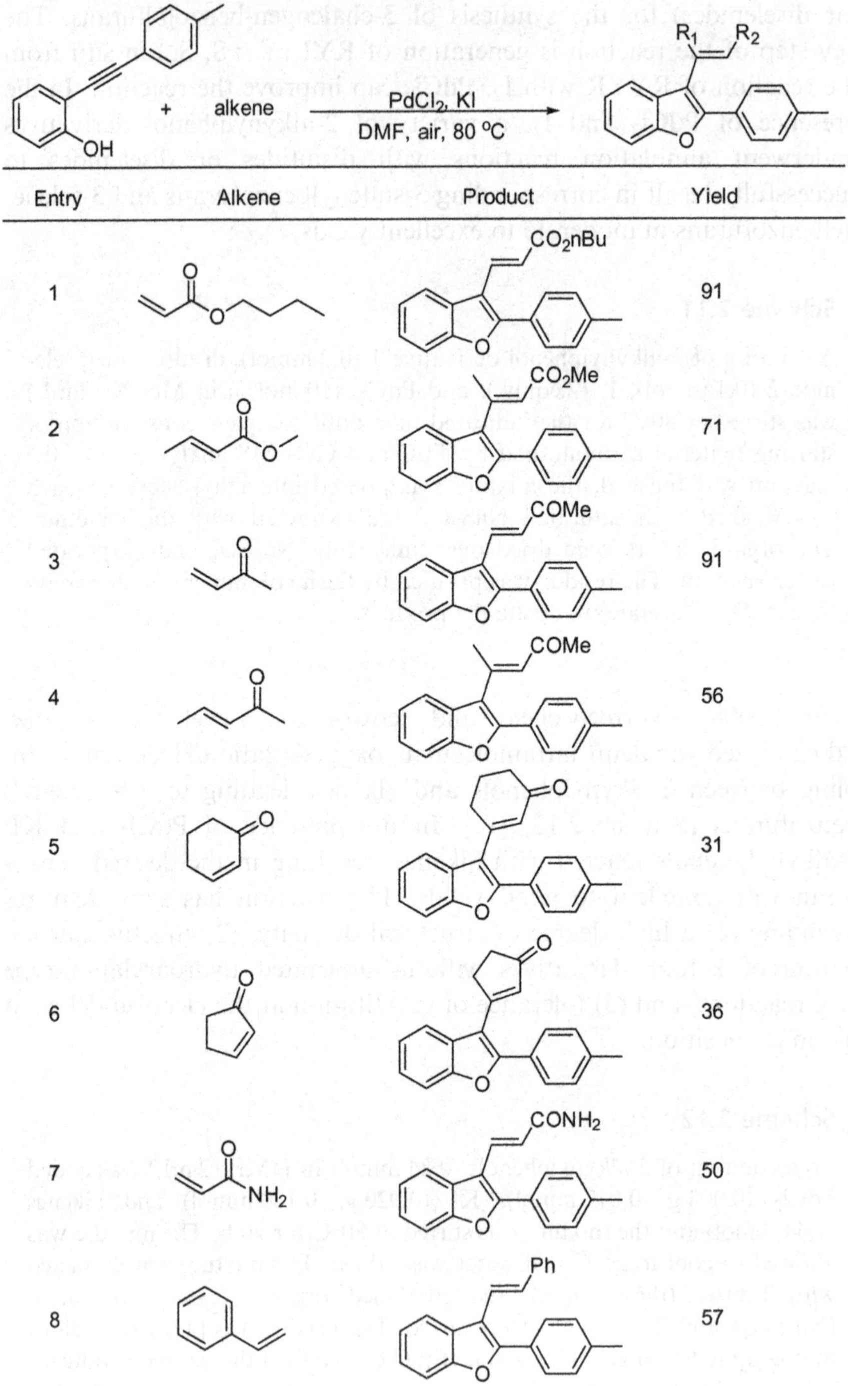

Entry	Alkene	Product	Yield
1		CO$_2$nBu	91
2		CO$_2$Me	71
3		COMe	91
4		COMe	56
5			31
6			36
7		CONH$_2$	50
8		Ph	57

Scheme 2.12 Palladium-catalyzed cyclization of 2-alkynylphenols and alkenes.

More recently, Sridhar Reddy and coworkers palladium-catalyzed tandem intramolecular oxy/amino-palladation/isocyanide insertion for the synthesis of α-benzofuranyl/indolyl-acetamides (Scheme 2.13) [12]. The reaction does not need oxidant or ligand to promote the cascade and can be carried out in open air. The reaction also includes isocyanide insertion.

In 2015, Chen and coworkers reported one-pot synthesis of 2,3-disubstituted dihydro-benzofurans and benzofurans via rhodium, palladium-catalyzed intramolecular C−H insertion reaction (Scheme 2.14) [13]. The reactions proceeded via rhodium-catalyzed ring-opening of N-sulfonyl-1,2,3-triazles, the intramolecular C_{sp3}−H insertion reaction of a-imino rhodium carbene, and subsequent palladium-catalyzed dehydrogenation. 2,3-Disubstituted hydrobenzofurans and benzofurans can be obtained in good to excellent yields in a one-pot manner.

The formation of 4 appears to proceed via the insertion of a-imino rhodium carbene toward a C−H bond adjacent to oxygen and subsequent isomerization of the double bond in the presence of a Pd/C catalyst. However, when the reaction was carried out under N_2 atmosphere instead of H_2, the isolated 2 was transformed into benzofuranyl imine 3 rather than 4 in the presence of Pd/C in CH_2Cl_2 or EtOH. This indicates that the transformation of 2−4 is not a simple isomerization process. Similarly, 3 could also be obtained from 1 via the one-pot, two-step process in N_2 atmosphere (Scheme 2.15).

In 2015, Valdes and coworkers reported Pd-catalyzed synthesis of benzo[b] fused carbo- and heterocycles through carbine migratory insertion (Scheme 2.16) [14]. This method is an original approach to substituted benzofurans, which are constructed by the formation of two C−C bonds on the same carbon atom. However, the yield of this reaction is low.

In 2016, Jiang and coworkers developed palladium-catalyzed cascade annulation/arylthiolation reaction in order to create functionalized 3-sulfenylbenzofuran derivatives in moderate to good yields from readily available 2-alkynylphenols and 2-alkynylamines in ionic liquids (Scheme 2.17) [15]. This protocol provides a valuable synthetic tool for the assembly of a wide range of 3-sulfenylbenzofuran derivatives. Another feature of this method is the employment of ionic liquids under mild reaction conditions.

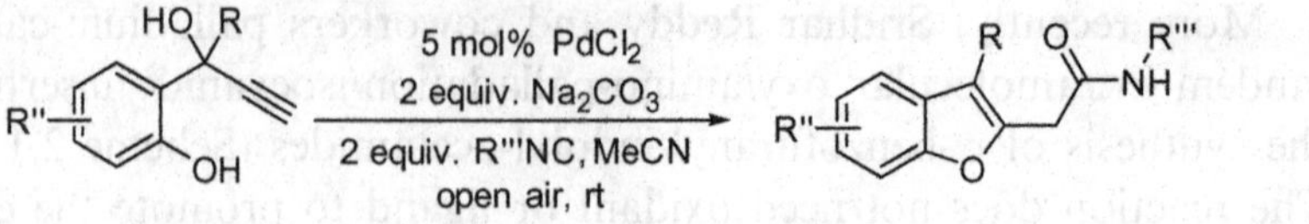

Entry	Product	Yield
1	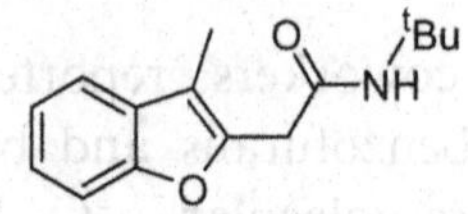	85%
2	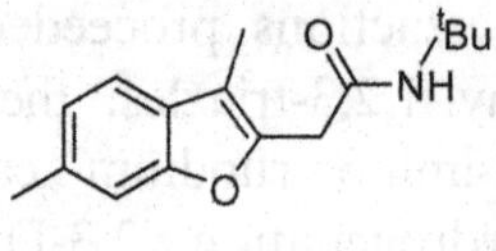	82%
3	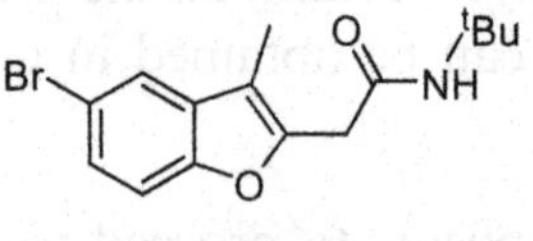	74%
4	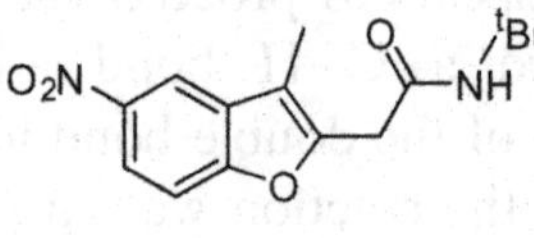	82%
5	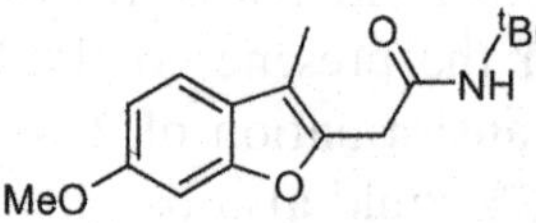	52%
6	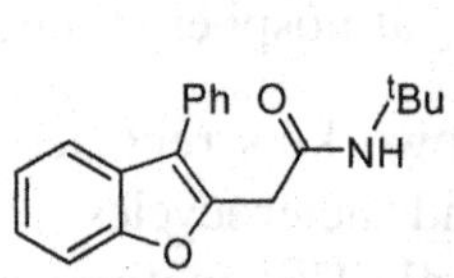	78%
7	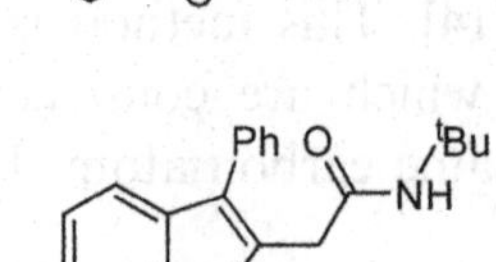	75%
8	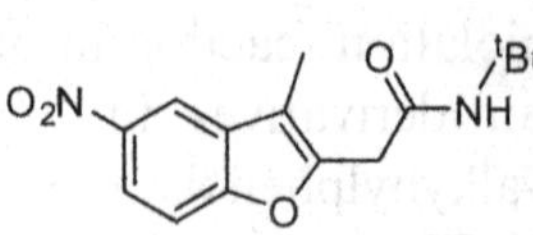	81%

Scheme 2.13 Synthesis of α-benzofuranylacetamides via oxypalladation and isocyanide insertion.

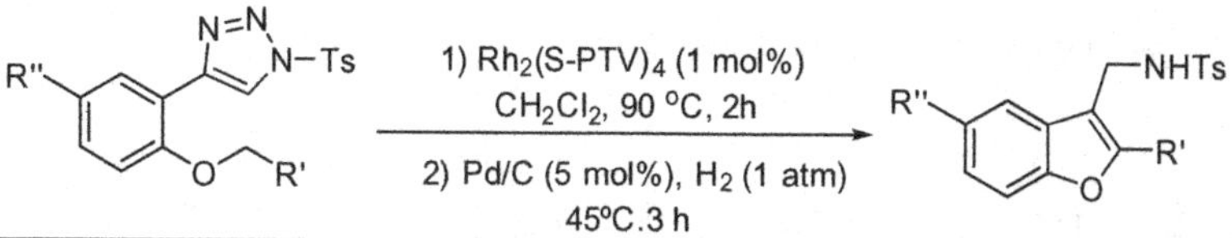

Entry	Product	Yield
1	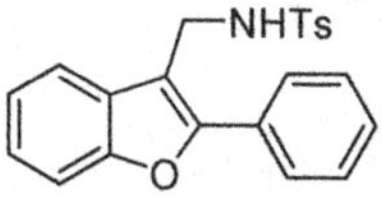	90%
2	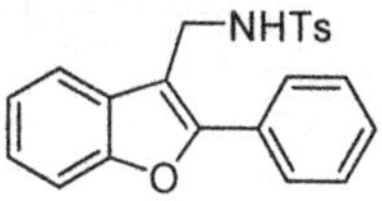	91%
3	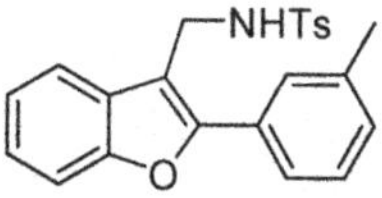	88%
4	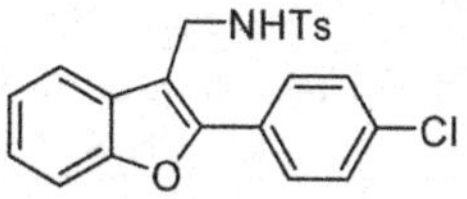	68%
5	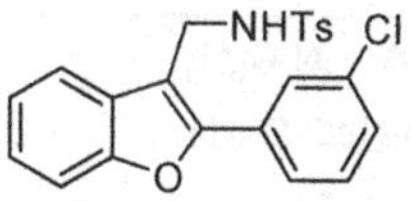	76%
6	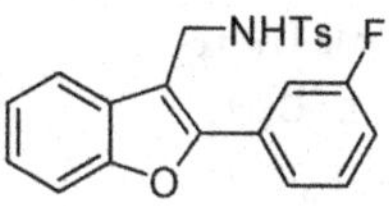	81%
7	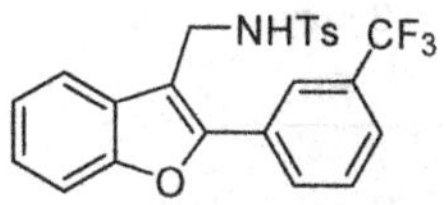	82%
8	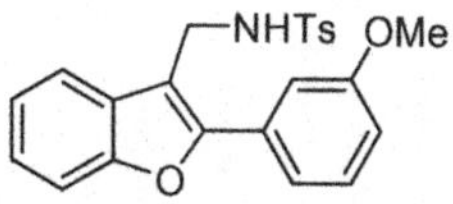	70%
9	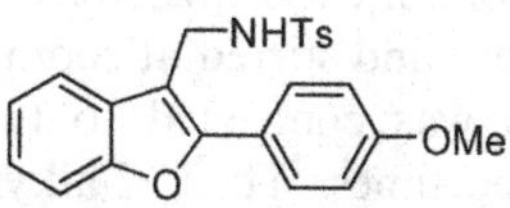	57%

Scheme 2.14 One-pot synthesis of benzofurans.

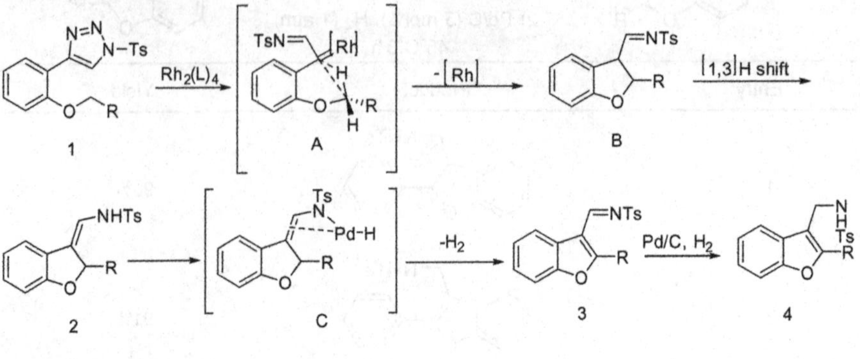

Scheme 2.15 Possible mechanism.

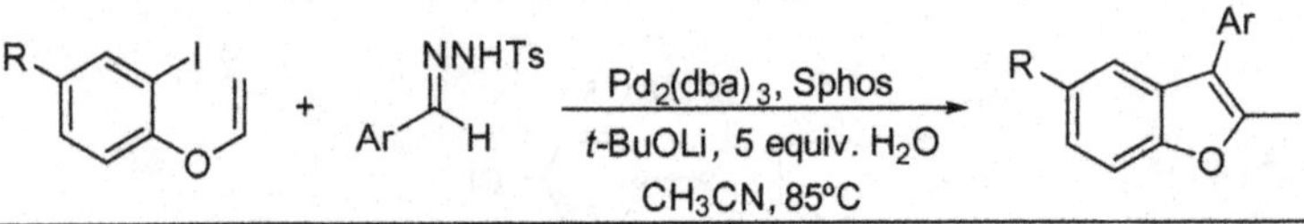

R	Ar	Yield
H	4-Me$_2$N-C$_6$H$_4$	50%
H	4-MeO-C$_6$H$_4$	77%
H	Ph	60%
H	4-F-C$_6$H$_4$	57%
H	4-Me-C$_6$H$_4$	59%
H	3-Cl-C$_6$H$_4$	58%
H	4-CF$_3$-C$_6$H$_4$	68%
H	4-CN-C$_6$H$_4$	73%
H	2-Furyl	40%
MeO	Ph	68%

Scheme 2.16 Palladium-catalyzed synthesis of benzofurans.

Scheme 2.17

Pd(TFA)$_2$ (5 mol%) and [Bmim]Cl (1 mL) were combined in an Schlenk tube equipped with a stir-bar and stirred at room temperature for 10 min. A balloon filled with N$_2$ was connected to the Schlenk tube via the side tube and purged three times. Then, 2-alkynylphenols (0.20 mmol), arylboronic acid (0.4 mmol), S$_8$ (0.60 mmol), CuI (0.40 mmol), Phen

(0.44 mmol), Ag_2CO_3 (0.4 mmol), and K_3PO_4 (0.4 mmol) were quickly added to the tube under N_2 atmosphere and stirred at 80°C for 8 h. After the reaction was completed, the N_2 gas was released carefully, and the reaction was quenched by water and extracted with CH_2Cl_2 three times. The combined organic layers were dried over anhydrous Na_2SO_4 and evaporated under vacuum. The desired products were obtained in the corresponding yields after purification by flash chromatography on silica gel with hexanes/ethyl acetate.

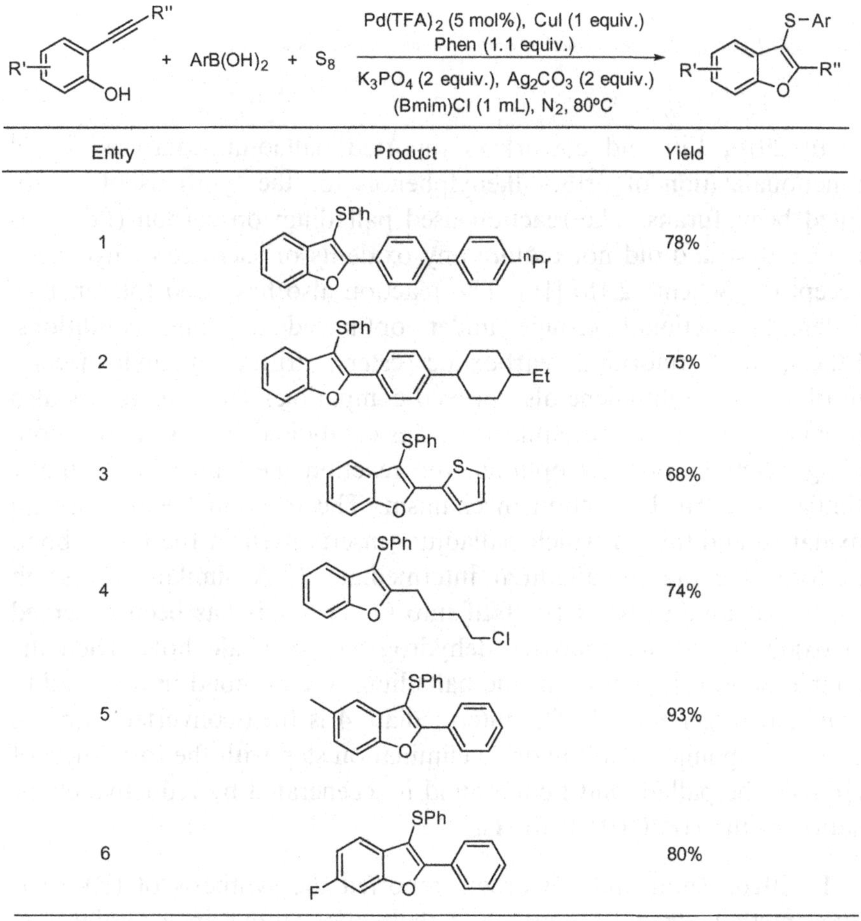

Entry	Product	Yield
1		78%
2		75%
3		68%
4		74%
5		93%
6		80%

Scheme 2.17 Palladium-catalyzed synthesis of 3-sulfenylbenzofurans.

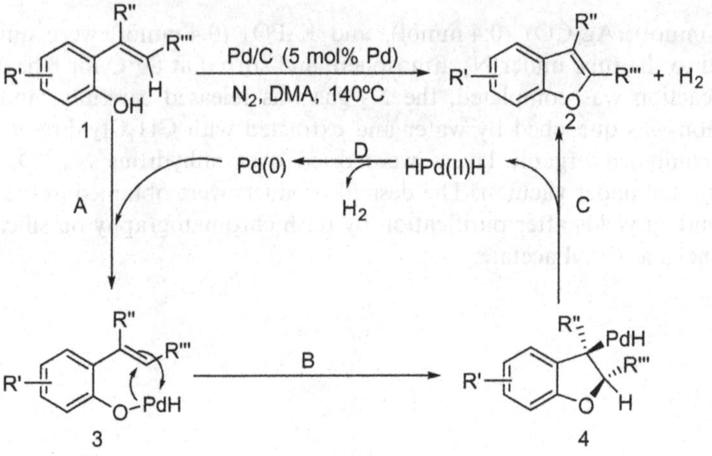

Scheme 2.18 Palladium-catalyzed synthesis of benzofurans and proposed mechanism.

In 2016, Liu and coworkers reported palladium-catalyzed C−H functionalization of ortho-alkenylphenols for the synthesis of substituted benzofurans. The reaction used palladium on carbon (Pd/C) as the catalyst and did not require any oxidants or sacrifice of hydrogen acceptors (Scheme 2.18) [16]. The reaction also has good tolerance of different functional groups under optimized coupling conditions. Ethers, aryl fluorides, carboxylic esters, trifluoromethylbenzenes, nitriles, and naphthalene also proved compatible. The researchers also provided an overall mechanism for this catalytic dehydrogenative coupling reaction. In their opinion, the reaction mechanism is formally similar to a Heck reaction mechanism. The reaction begins with an oxidative addition in which palladium inserts itself in the O−H bond to form the organopalladium intermediate 3. A similar process in which palladium (0) inserts itself into O−H bonds has been observed in oxidant- and acceptor-free dehydrogenation of alcohols. Then the double bond inserts itself in the palladium-oxygen bond in a syn addition step to generate 4. The intermediate 4 is then converted into the target compound by a β-hydride elimination step with the formation of HPdH. The palladium(0) compound is regenerated by reductive elimination of the HPd(II)H with H_2.

In 2016, Zhou and coworkers reported the synthesis of (Z)-α-trifluoromethyl alkenyl triflates and their applications in a straightforward synthesis of trifluoromethylated benzofurans (Scheme 2.19) [17].

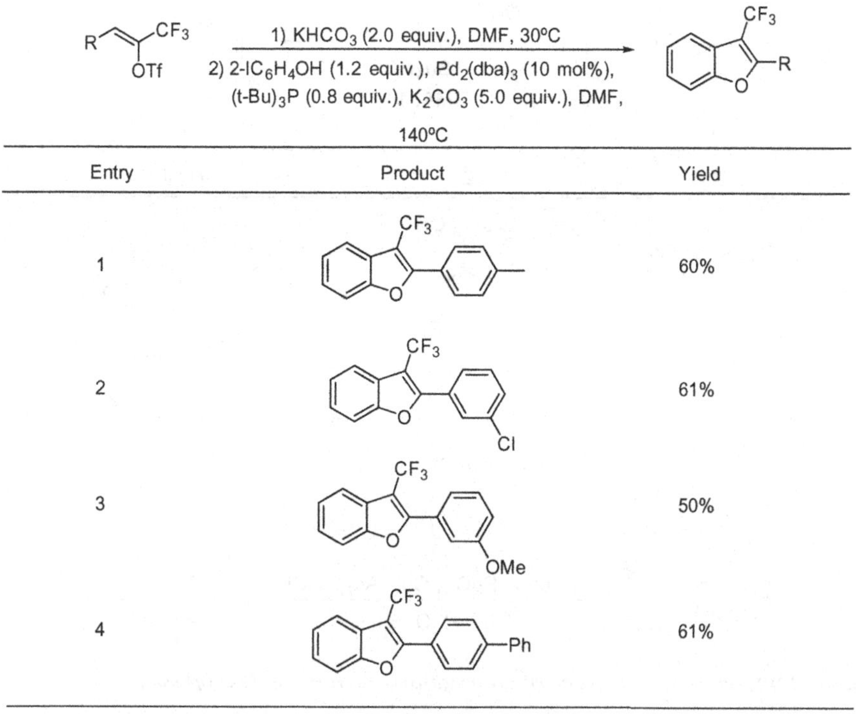

Scheme 2.19 Synthesis of 3-trifluoromethylbenzofuran.

The reaction used Pd(OAc)$_2$ as the catalyst, X-Phos as the ligand, and Cs$_2$CO$_3$ as the base. Their method can also be used for the synthesis of 3-trifluoromethylbenzofuran, which is an essential structural motif in biologically active compounds.

In 2011, Li and coworkers reported palladium-catalyzed selective heck-type diarylation of allylic esters with aryl halides. This methodology can also be applied in constructing the indole and benzofuran skeletons (Scheme 2.20) [18]. Both 2-iodophenol and 2-bromophenol are also suitable substrates, providing the desired benzofuran in satisfactory yields.

In 1998, Bumagin and coworkers developed Pd^{2+}-catalyzed oxidative cyclization of 2-allylphenols for synthesis of benzofurans (Scheme 2.21) [19]. Oxidative cyclization of 2-allylphenols can be performed easily and under mild conditions using palladium dichloride as a catalyst, Cu(OAc)$_2$–LiCl as a reoxidant, and aqueous

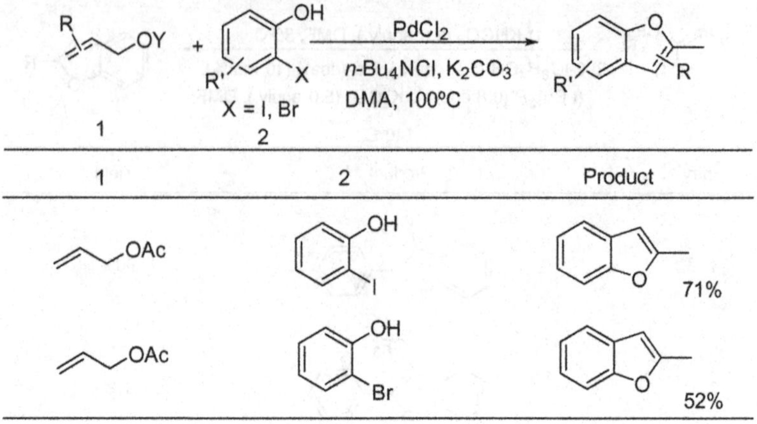

Scheme 2.20 Palladium-catalyzed synthesis of benzofuran.

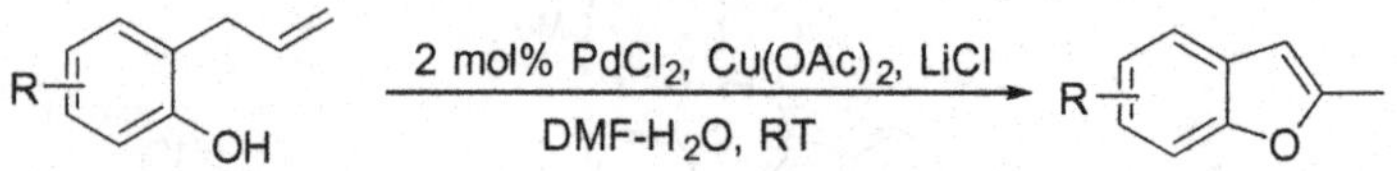

Scheme 2.21 Synthesis of benzofurans via Pd^{2+}-catalyzed oxidative cyclization of 2-allylphenols.

dimetyl-formamide as a solvent. A number of functionalized 2-methyl-bezofurans were obtained in high yields.

In 2010, Correia and coworkers developed an intramolecular Heck−Matsuda reaction and applied it in the syntheses of benzofurans and indoles (Scheme 2.22) [20]. The reaction has a wide scope of substrates. Both electron-donating and/or electron-withdrawing groups at the aromatic position can be tolerated. However, electron-rich, -poor and -neutral substituents resulted in Heck adducts in lower yields, compared to those obtained for an unsubstituted benzofuran.

In 2011, Lu and coworkers reported the synthesis of 1-benzoxepine derivatives via cationic palladium-catalyzed [5 + 2] annulation reaction of 2-acylmethoxyarylboronic acids with allenoates (Scheme 2.23) [21]. This annulation involves the intramolecular nucleophilic addition to ketones without the formation of π-allylpalladium species.

In 2015, Stahl and coworkers reported a palladium-catalyzed aerobic oxidative dehydrogenation system that can be used for synthesis of benzofurans (Scheme 2.24) [22]. The reaction used O_2 as the

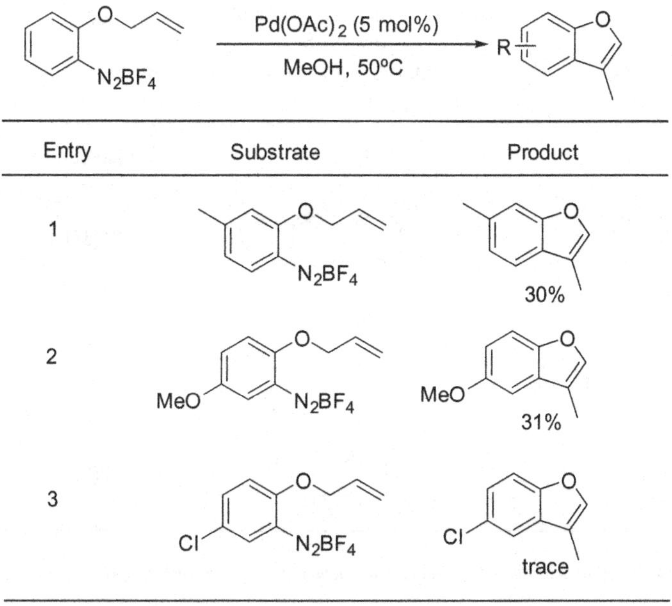

Entry	Substrate	Product
1		30%
2		31%
3		trace

Scheme 2.22 Synthesis of benzofurans via Pd^{2+}-catalyzed cyclization of the aryl o-allylether.

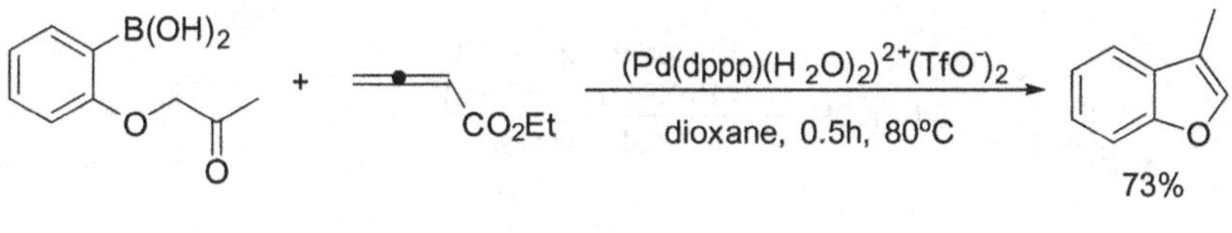

Scheme 2.23 Synthesis of benzofurans via annulation reaction of the 2-formylphenylboronic acids with substituted allenoates.

oxidant, which enables efficient synthesis of substituted arene derivatives and shows good functional group tolerance.

In 2011, Sanz and coworkers reported a Pd-catalyzed Buchwald protocol in an attempt to synthesize the desired dimethoxybenzo[b]furan derivatives (Scheme 2.25) [23]. The reaction of the previously prepared o-haloaryl alkynes with KOH, in the presence of catalytic amounts of $Pd_2(dba)_3$ and t-BuXPhos at 100°C in a 1:1 mixture of H_2O:1,4-dioxane, gave rise to the corresponding benzo[b]furan derivatives in moderate to good yields.

In 2013, Chen and coworkers developed a new synthesis of 2-arylbenzofurans by palladium-catalyzed reaction of arylboronic acid with

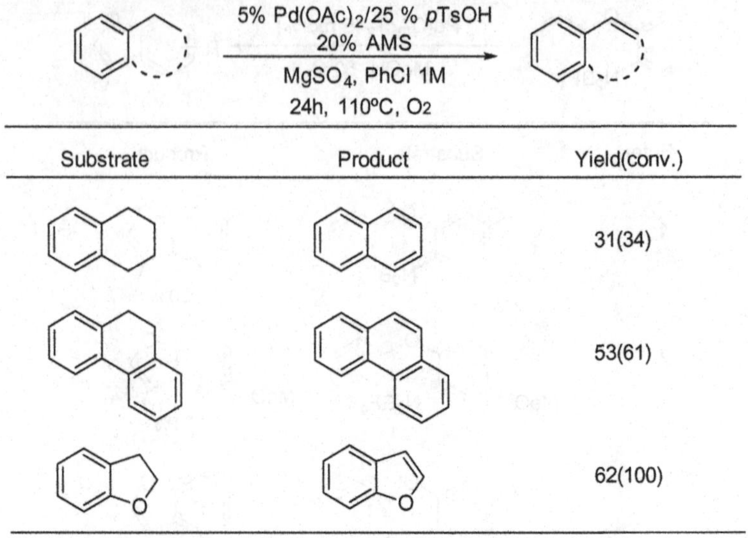

Substrate	Product	Yield(conv.)
		31(34)
		53(61)
		62(100)

Scheme 2.24 Synthesis of benzofurans via annulation reaction of the 2-formylphenylboronic acids with substituted allenoates.

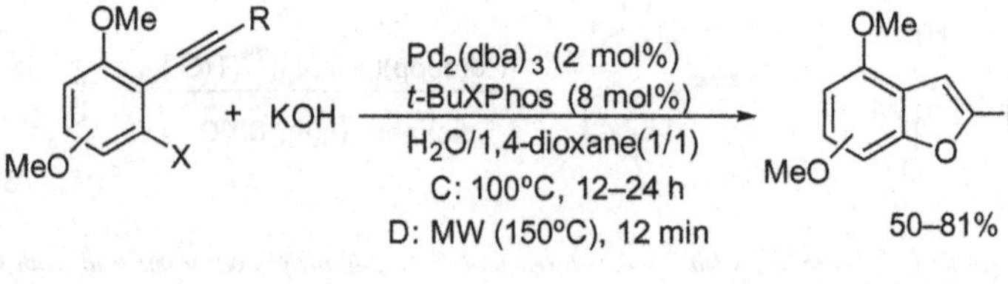

Scheme 2.25 Synthesis of benzofurans.

aliphatic nitriles (Scheme 2.26) [24]. The reactions used Pd(OAc)$_2$ (5 mol%) as the catalyst, bpy (10 mol%) as the ligand, TFA as the additive, and THF-H$_2$O as the solvent. The electronic properties of the groups on the phenyl ring of the benzoboronic acid had some effect on the yield of the reaction. Arylboronic acids, which have electron-donating groups at the *para* position, have higher yield.

In 2013, the same group also developed a new method for synthesis of benzofurans from potassium aryltrifluoroborates and hydroxyphenylacetonitriles (Scheme 2.27) [25]. The reaction used Pd(CF$_3$CO$_2$)$_2$ as the catalyst, bpy (10 mol%) as the ligand, TFA (10 equiv.) as the additive, and THF-H$_2$O as the solvent.

$$ArB(OH)_2 \quad + \quad R \overset{\text{CN}}{\underset{\text{OH}}{\bigcirc}} \quad \xrightarrow[\substack{bpy,TFA,\ THF-H_2O, \\ N_2,\ 80\ ^\circ C}]{Pd(OAc)_2\ (5\ mol\%)} \quad R \overset{\bigcirc}{\underset{O}{\bigcirc}} - Ar$$

Entry	R	Ar	Yield[b] (%)
1	H	Ph	94
2	H	4-MeC$_6$H$_4$	91
3	H	4-MeOC$_6$H$_4$	92
4	H	4-FC$_6$H$_4$	86
5	H	4-ClC$_6$H$_4$	84
6	H	4-BrC$_6$H$_4$	81
7	H	4-PhC$_6$H$_4$	85
8	4-Me	Ph	91
9	3-MeO	Ph	89
10	4-Cl	Ph	81
11	4-Br	Ph	80

Scheme 2.26 Palladium-catalyzed synthesis of 2-arylbenzofurans.

In 2010, Manabe and coworkers reported palladium-catalyzed synthesis of bentofuran from 2-chlorophenols (Scheme 2.28) [26]. This method used a bifunctional ligand **L1** for cross-coupling of chloroarenes, and provided a new example of benzofuran synthesis from 2-chlorophenol.

In 2008, Burch and coworkers developed a palladium-catalyzed one-pot synthesis of benzofuran from *o*-bromophenols (Scheme 2.29) [27]. The reaction used Pd(OAc)$_2$ as the catalyst, *rac*-DTBPB as the ligand, and NaO^{t}Bu as the additive. Then, treatment of the reaction mixture with a 1:1 mixture of CH$_2$Cl$_2$ and trifluoroacetic acid provided the benzofurans.

In 2008, Li and coworkers developed a palladium/copper cocatalyzed cross-coupling reaction of terminal alkynes with (*E*)-2,3-diiodoalkenes for the synthesis of benzofuran (Scheme 2.30) [28]. The reaction used Pd(OAc)$_2$ and CuI as the catalyst, Et$_3$N as the base, and THF as the solvent.

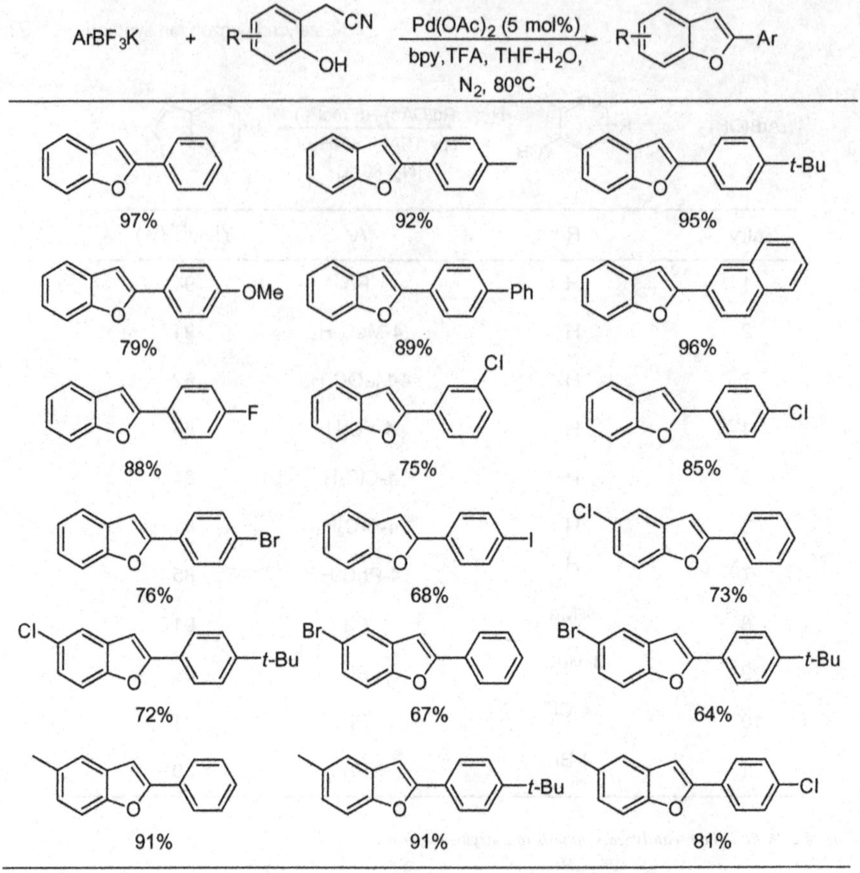

$ArBF_3K$ + (2-hydroxyphenylacetonitrile) $\xrightarrow[\substack{bpy, TFA, THF-H_2O, \\ N_2, 80°C}]{Pd(OAc)_2 \ (5 \ mol\%)}$ (2-aryl benzofuran)

97%   92%   95%

79%   89%   96%

88%   75%   85%

76%   68%   73%

72%   67%   64%

91%   91%   81%

Scheme 2.27 Synthesis of benzofurans from the potassium aryltrifluoroborates to 2-hydroxyphenylacetonitriles.

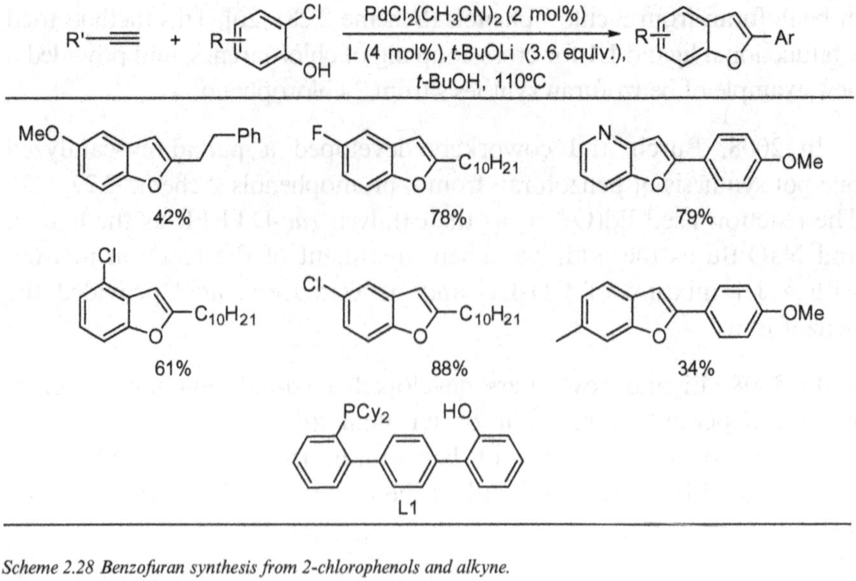

R'—≡ + (2-chlorophenol) $\xrightarrow[\substack{L \ (4 \ mol\%), t\text{-BuOLi} \ (3.6 \ equiv.), \\ t\text{-BuOH}, \ 110°C}]{PdCl_2(CH_3CN)_2 \ (2 \ mol\%)}$ (2-aryl benzofuran)

42%   78%   79%

61%   88%   34%

L1

Scheme 2.28 Benzofuran synthesis from 2-chlorophenols and alkyne.

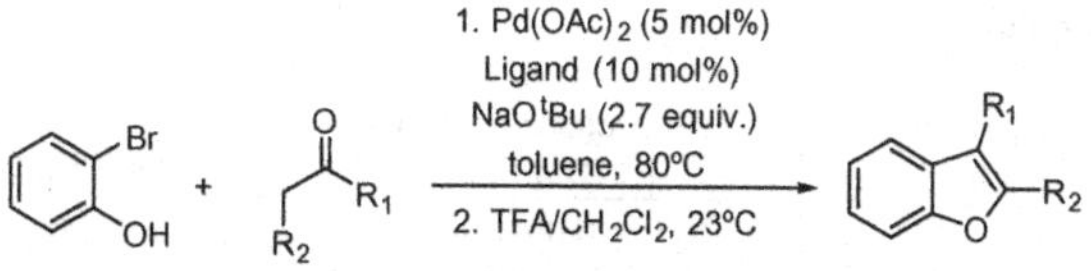
Br
OH
+
O
R1
R2
1. Pd(OAc)2 (5 mol%)
Ligand (10 mol%)
NaOtBu (2.7 equiv.)
toluene, 80°C
2. TFA/CH2Cl2, 23°C
R1
R2
O

Entry	Substrate	Product	Yield [%]
1	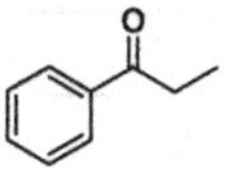	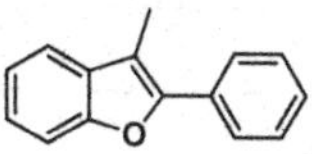	76
2	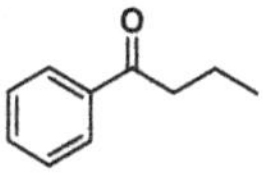	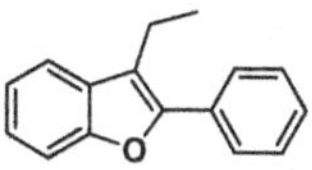	59
3	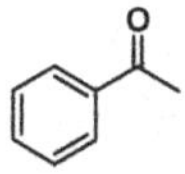	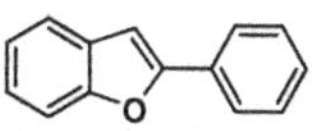	73
4	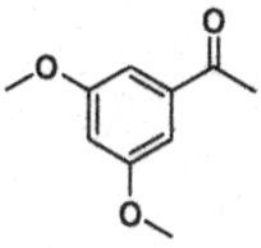	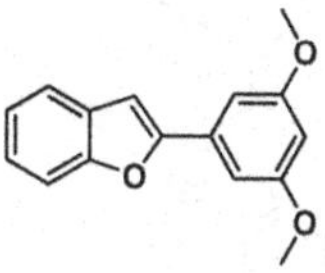	65
5	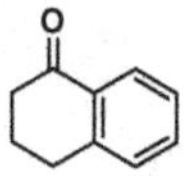	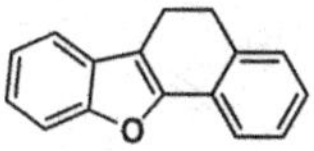	84
6	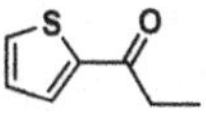	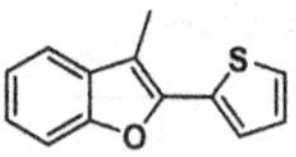	70
7	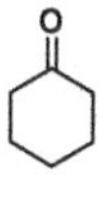	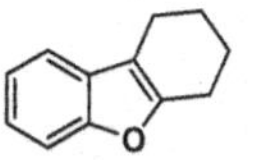	41
8	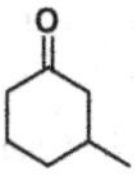	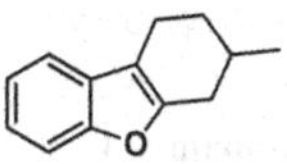	75
9	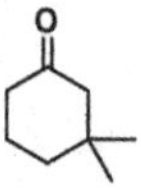	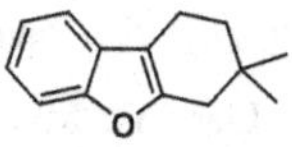	84

Scheme 2.29 One-pot synthesis of benzofurans.

THF as the solvent.

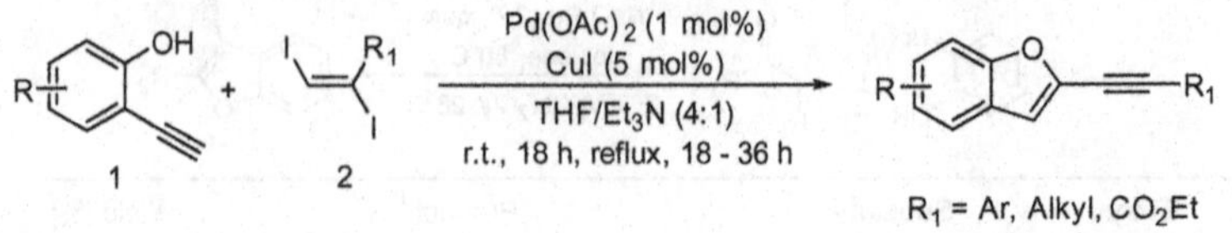

R_1 = Ar, Alkyl, CO_2Et

Entry	1	2	Yield (%)
1	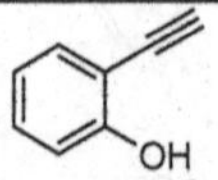	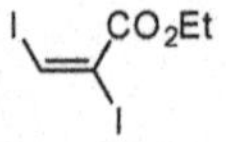 CO_2Et	33
2	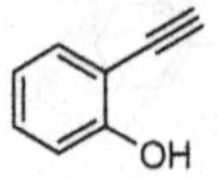	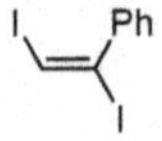 Ph	63
3	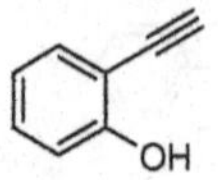	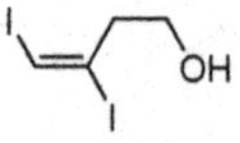 OH	58
4	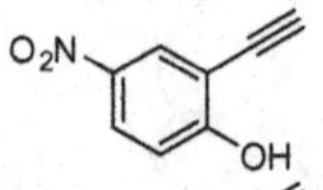 O_2N	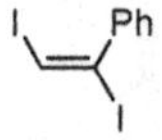 Ph	91
5	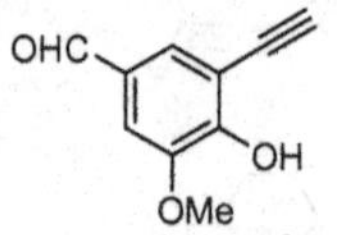 OHC / OMe	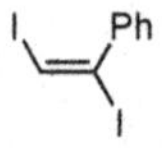 Ph	36
6	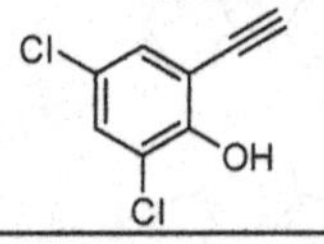 Cl / Cl	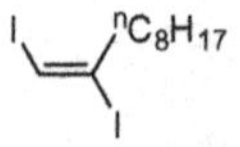 $^nC_8H_{17}$	57

Scheme 2.30 One-pot synthesis of 2-ethynylbenzofurans.

REFERENCES

[1] Arcadi, A.; Cacchi, S.; Del Rosario, M.; Fabrizi, G.; Marinelli, F. *J. Org. Chem.* **1996**, *61*, 9280−9288.

[2] Kundu, N. G.; Pal, M.; Mahanty, J. S.; Dasgupta, S. K. *J. Chem. Soc., Chem. Commun.* **1992**, *1*, 41−42.

[3] Nan, Y.; Miao, H.; Yang, Z. *Org. Lett.* **2000**, *2*, 297−299.

[4] Hu, Y.; Nawoschik, K. J.; Liao, Y.; Ma, J.; Fathi, R.; Yang, Z. *J. Org. Chem.* **2004**, *69*, 2235−2239.

[5] Willis, M. C.; Taylor, D.; Gillmore, A. T. *Org. Lett.* **2004**, *6*, 4755−4757.

[6] Xie, X.; Chen, B.; Lu, J.; Han, J.; She, X.; Pan, X. *Tetrahedron Lett.* **2004**, *45*, 6235−6237.

[7] Yue, W. S.; Li, J. J. *Org. Lett.* **2002**, *4*, 2201−2203.

[8] Nakamura, I.; Mizushima, Y.; Yamamoto, Y. *J. Am. Chem. Soc.* **2005**, *127*, 15022−15023.

[9] Liang, Y.; Tang, S.; Zhang, X.-D.; Mao, L.-Q.; Xie, Y.-X.; Li, J.-H. *Org. Lett.* **2006**, *8*, 3017−3020.

[10] Du, H.-A.; Zhang, X.-G.; Tang, R.-Y.; Li, J.-H. *J. Org. Chem.* **2009**, *74*, 7844–7848.

[11] Martínez, C.; Álvarez, R.; Aurrecoechea, J. M. *Org. Lett.* **2009**, *11*, 1083–1086.

[12] Thirupathi, N.; Hari Babu, M.; Dwivedi, V.; Kant, R.; Sridhar Reddy, M. *Org. Lett.* **2014**, *16*, 2908–2911.

[13] Ma, X.; Wu, F.; Yi, X.; Wang, H.; Chen, W. *Chem. Commun.* **2015**, *51*, 6862–6865.

[14] Paraja, M.; Carmen Perez-Aguilar, M.; Valdes, C. *Chem. Commun.* **2015**, *51*, 16241–16243.

[15] Li, J.; Li, C.; Yang, S.; An, Y.; Wu, W.; Jiang, H. *J. Org. Chem.* **2016**, *81*, 2875–2887.

[16] Yang, D.; Zhu, Y.; Yang, N.; Jiang, Q.; Liu, R. *Adv. Synth. Catal.* **2016**, *358*, 1731–1735.

[17] Zhao, Y.; Zhou, Y.; Liu, J.; Yang, D.; Tao, L.; Liu, Y., et al. *J. Org. Chem.* **2016**, *81*, 4797–4806.

[18] Liu, Y.; Yao, B.; Deng, C.-L.; Tang, R.-Y.; Zhang, X.-G.; Li, J.-H. *Org. Lett.* **2011**, *13*, 1126–1129.

[19] Roshchin, A. I.; Kel'chevski, S. M.; Bumagin, N. A. *J. Organomet. Chem.* **1998**, *560*, 163–167.

[20] Siqueira, F. A.; Taylor, J. G.; Correia, C. R. D. *Tetrahedron Lett.* **2010**, *51*, 2102–2105.

[21] Yu, X.; Lu, X. *J. Org. Chem.* **2011**, *76*, 6350–6355.

[22] Iosub, A. V.; Stahl, S. S. *J. Am. Chem. Soc.* **2015**, *137*, 3454–3457.

[23] Guilarte, V.; Castroviejo, M. P.; Álvarez, E.; Sanz, R. *Beilstein J. Org. Chem.* **2011**, *7*, 1255–1260.

[24] Wang, X.; Wang, X.; Liu, M.; Ding, J.; Chen, J.; Wu, H. *Synthesis* **2013**, *45*, 2241–2244.

[25] Wang, X.; Liu, M.; Xu, L.; Wang, Q.; Chen, J.; Ding, J., et al. *J. Org. Chem.* **2013**, *78*, 5273–5281.

[26] Wang, J.-R.; Manabe, K. *J. Org. Chem.* **2010**, *75*, 5340–5342.

[27] Eidamshaus, C.; Burch, J. D. *Org. Lett.* **2008**, *10*, 4211–4214.

[28] Liang, Y.; Tao, L.-M.; Zhang, Y.-H.; Li, J.-H. *Synthesis* **2008**, 3988–3994.

CHAPTER 3

Cu-Catalyzed Benzofuran Synthesis

In 2005, Chen and coworker reported a CuI-catalyzed ring closure of 2-haloaromatic ketones for the synthesis of benzo[b]furan (Scheme 3.1) [1]. The catalyzed protocol can tolerate various functional

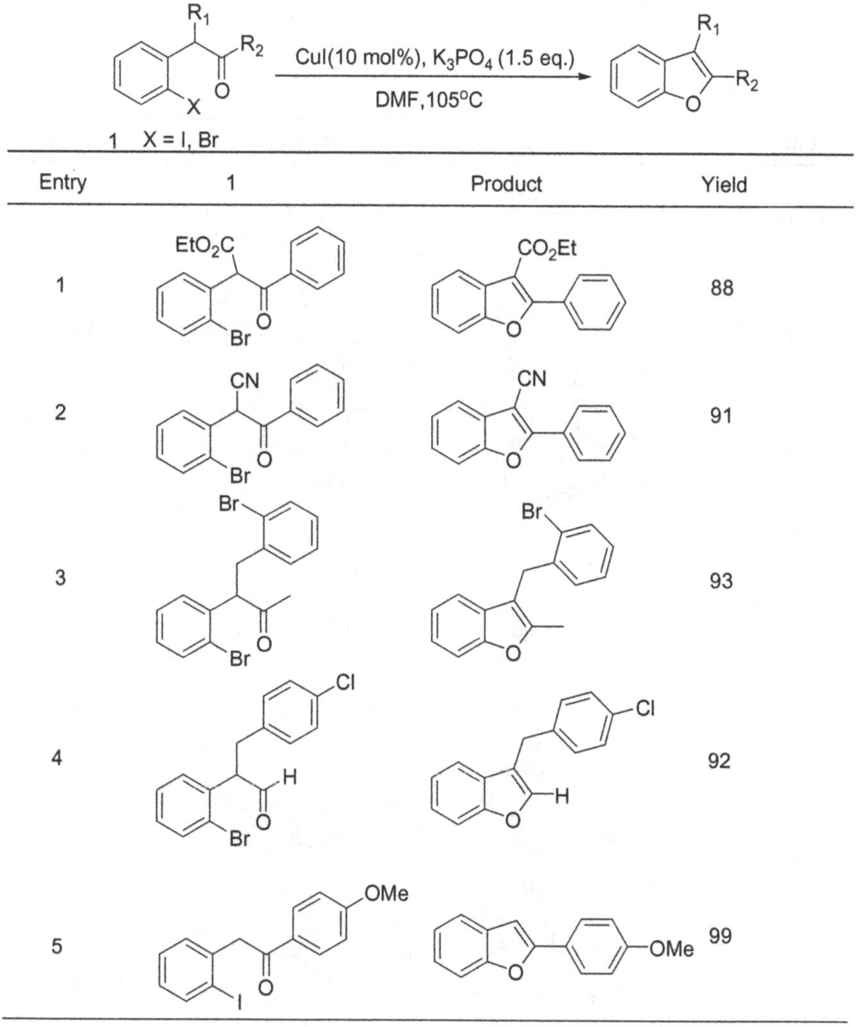

Entry	1	Product	Yield
1			88
2			91
3			93
4			92
5			99

Scheme 3.1 Synthesis of benzo[b]furans via CuI-catalyzed ring closure.

Transition Metal-Catalyzed Benzofuran Synthesis. DOI: http://dx.doi.org/10.1016/B978-0-12-809377-1.00003-6

groups. The researchers also suggested the reaction occurs through the intramolecular $S_{RN}1$ mechanism.

In 2006, SanMartin and coworker reported copper-catalyzed synthesis of benzo[*b*]-furan derivatives in water (Scheme 3.2) [2]. The reaction involved a copper–TMEDA complex and used ketone derivatives as the starting substances. One of the advantages of this reaction is

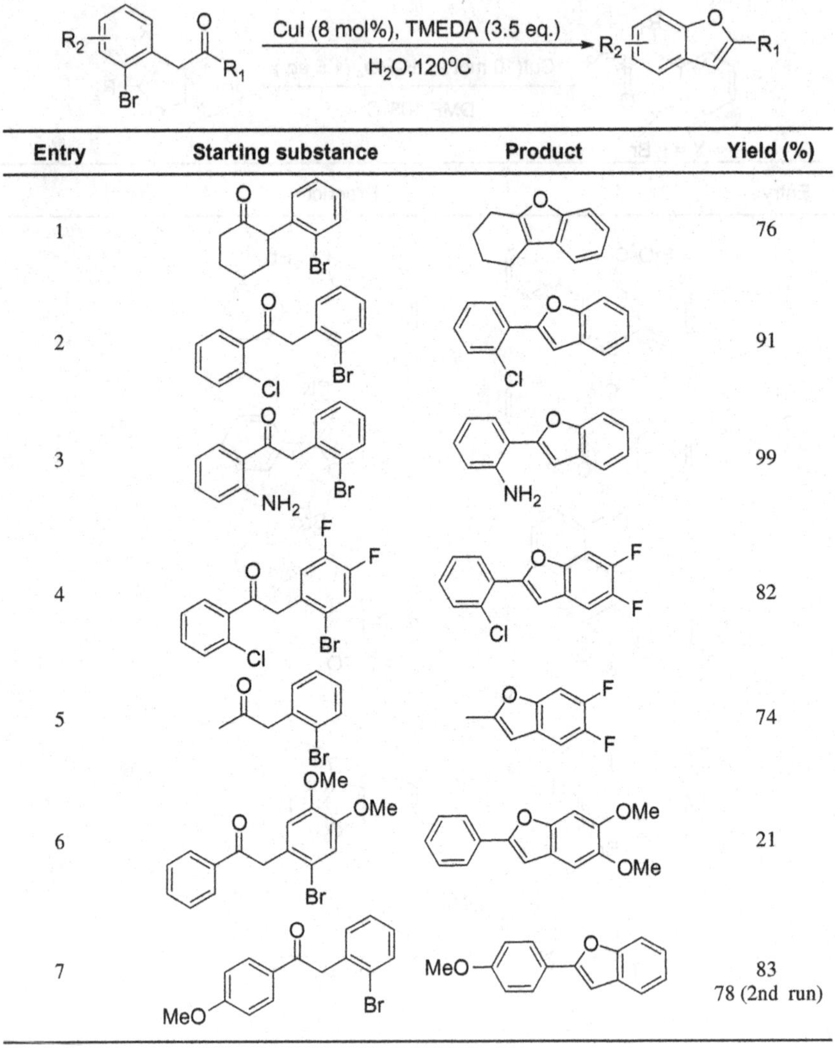
CuI (8 mol%), TMEDA (3.5 eq.)
H2O,120°C
Entry
Starting substance
Product
Yield (%)
1
76
2
91
3
99
4
82
5
74
6
21
7
83
78 (2nd run)

Scheme 3.2 Copper-catalyzed straightforward synthesis of benzo[b]furan derivatives in neat water.

using water as the solvent. The benefits of using water are clear in terms of lack of toxicity, safety, and cost.

Scheme 3.2

A schlenk tube was charged with deoxybenzoin (0.16 mmol), CuI (2.6 mg, 0.0136 mmol), TMEDA (85 µL, 0.56 mmol), and water (1.9 mL). Then the tube was sealed under a positive pressure of argon and the obtained green solution was heated overnight at 120°C. The product was extracted from the aqueous layer with dichloromethane (3 × 5 mL), dried, and concentrated in vacuo. The crude mixture was then purified by flash chromatography (40% hexane/CH$_2$Cl$_2$) to create benzofuran.

In 2007, Lautens and coworker developed a general method of benzofuran synthesis via Cu- and Pd-catalyzed cross-coupling (Scheme 3.3) [3].

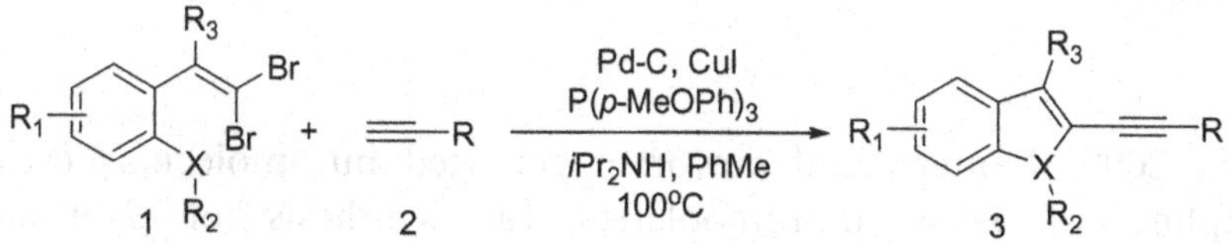

Entry	Alkyne	Product/yield (%)
1	≡—C$_6$H$_{13}$	80
2	≡—TMS	47
3	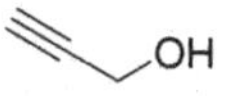 ⟋⟍OH	49
4	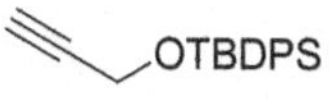 ⟋⟍OTBDPS	80
5	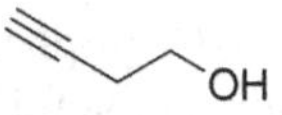 ⟋⟍⟍OH	63
6	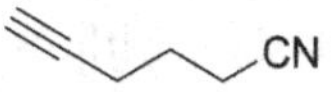 ⟋⟍⟍CN	76
7	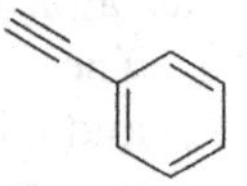 ⟋⟍C$_6$H$_5$	71
8	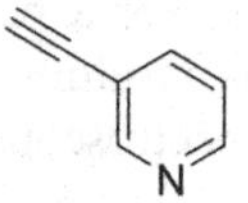 ⟋⟍pyridyl	61

Scheme 3.3 Copper-catalyzed straightforward synthesis of benzo[b]furan derivatives in neat water.

The reaction used Pd/C- and CuI as the catalyst and used gem-dibromovinyl substrates and terminal alkynes as the substances. Both gem-dibromovinyl substrates and terminal alkynes showed good tolerance on different groups.

Scheme 3.3

A carousel reaction tube (24×150 mm) was charged with 1 (0.41 mmol), 10% Pd-C (4.4 mg, 0.0041 mmol, 1 mol%), P(p-MeOPh)$_3$ (5.8 mg, 0.017 mmol, 4 mol%), and CuI (1.6 mg, 0.0083 mmol, 2 mol%), and was evacuated and purged with argon three times. Toluene (2 mL, degassed), H$_2$O (1 mL, degassed), iPr$_2$NH (145 μL, 1.03 mmol), and 2 (0.75 mmol) were added to this mixture and then heated to 100°C with stirring for 12 h. The reaction mixture was then cooled to room temperature and H$_2$O (10 mL) added. The mixture was extracted with Ethyl acetate (EtOAc) (2×15 mL) and combined extracts were washed with salt. NH$_4$Cl and brine, then dried, and solvent was removed in vacuo. The resulting crude material was purified by flash chromatography eluting with 2% EtOAc in hexane to create the product.

In 2009, Lautens and coworker reported intramolecular cross-coupling of gem- dibromoolefins for synthesis of 2-bromo-benzofused heterocycles (Scheme 3.4) [4]. Highly useful halogenated benzofurans are prepared from gem-dibromoolefins using a mild, ligand-free copper catalyzed cross-coupling procedure. In this process purification by flash chromatography is unnecessary, making this method highly efficient.

Scheme 3.4

A 0.5–2 mL microwave-reaction vial equipped with a magnetic stir bar was added to the requisite gem-dibromoolefin (1 eq.), CuI (5 mol%), and K$_3$PO$_4$ (2 eq.). The flask was flushed with argon for 5 min, after which THF (1 mL per 0.2 mmol olefin) was added and the vial sealed and placed in a preheated oil bath at 80°C. The vial was stirred for 6 h, after which it was removed from the oil bath and allowed to cool to room temperature. The contents were filtered over a pad of silica gel and washed with copious amounts of Et$_2$O. The resulting solution was concentrated under reduced pressure to create a spectroscopically pure product.

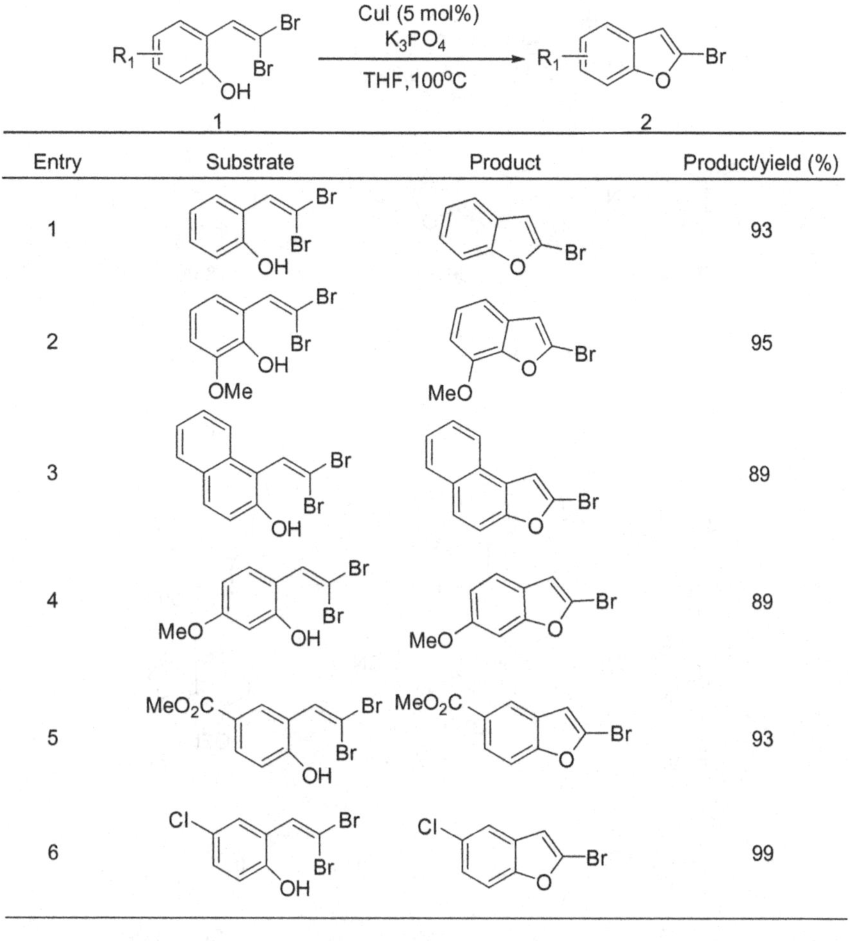

Scheme 3.4 Synthesis of benzofurans by intramolecular Ullmann coupling.

In 2012, Wang and coworker developed a highly efficient one-pot procedure by reaction of 2-(gem-dibromovinyl)phenols-(thiophenols) with $K_4Fe(CN)_6$ to 2-cyanobenzofurans(thiophenes) (Scheme 3.5) [5]. In the presence of $CuI/Na_2CO_3-Pd(OAc)_2/PPh_3$ in N,N-dimethylformamide (DMF), the reaction of 2-(gem-dibromovinyl) phenols and 2-(gem-dibromovinyl)thiophenols with $K_4Fe(CN)_6$, as a nontoxic and user-friendly cyanating reagent, generated the corresponding 2-cyanobenzofurans with good yields.

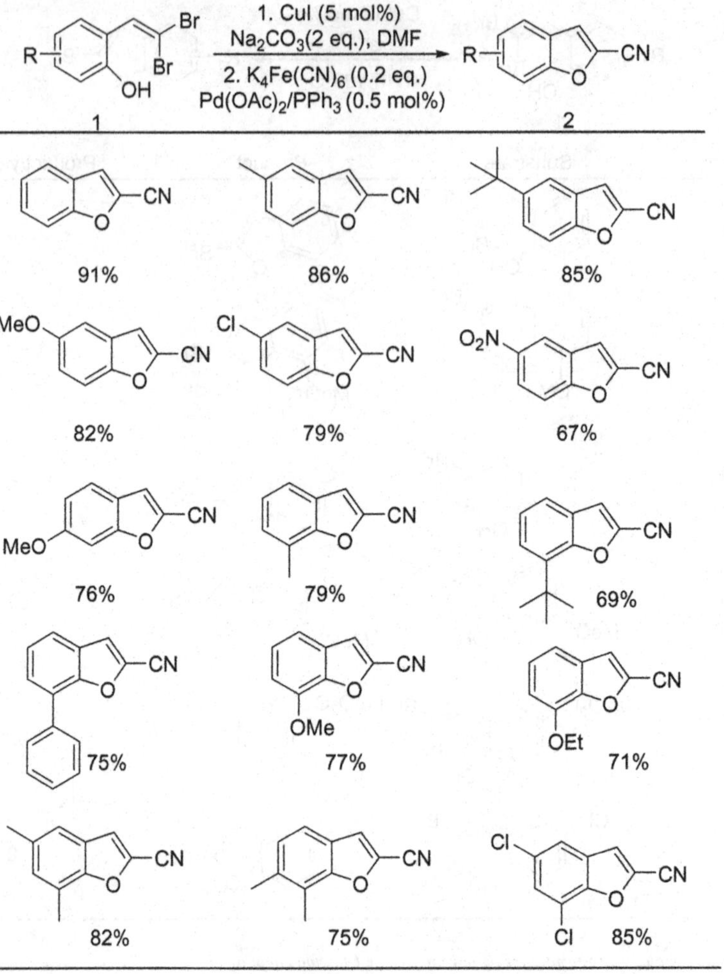

Scheme 3.5 *CulPd-catalyzed one-pot reactions of 2-(gem-dibromovinyl)phenols(thiophenols) with $K_4Fe(CN)_6$.*

Scheme 3.5

A sealable reaction tube equipped with a magnetic stirrer bar was charged with gem-dibromovinyl substrate (1.0 mmol), CuI (0.10 mmol), Na_2CO_3 (2.0 mmol), and DMF (2.0 mL). The rubber septum was then replaced by a Teflon-coated screw cap, and the reaction vessel placed in an oil bath at 80°C. After stirring of the mixture at this temperature for 6 h, it was cooled to room temperature and $K_4Fe(CN)_6$ (0.20 mmol), Pd (OAc)$_2$ (0.01 mmol), and PPh$_3$ (0.02 mmol) were added to the reaction system. Then the reaction vessel was placed in an oil bath at 120°C for 6 h. It was cooled to room temperature after the reaction and diluted with ethyl acetate, washed with water and brine, and dried with Mg_2SO_4. After the solvent was removed under reduced pressure, the residue was purified by column chromatography on silica gel (eluant: petroleum ether) to create the corresponding product.

In 2013, Wang and coworker developed trace amount Cu (ppm)-catalyzed intramolecular cyclization of 2-(gem-dibromovinyl)-phenols (thiophenols) to 2-bromo-benzofurans (thiophenes) (Scheme 3.6) [6]. 2-bromobenzofurans (thiophenes) can be obtained in the presence of a

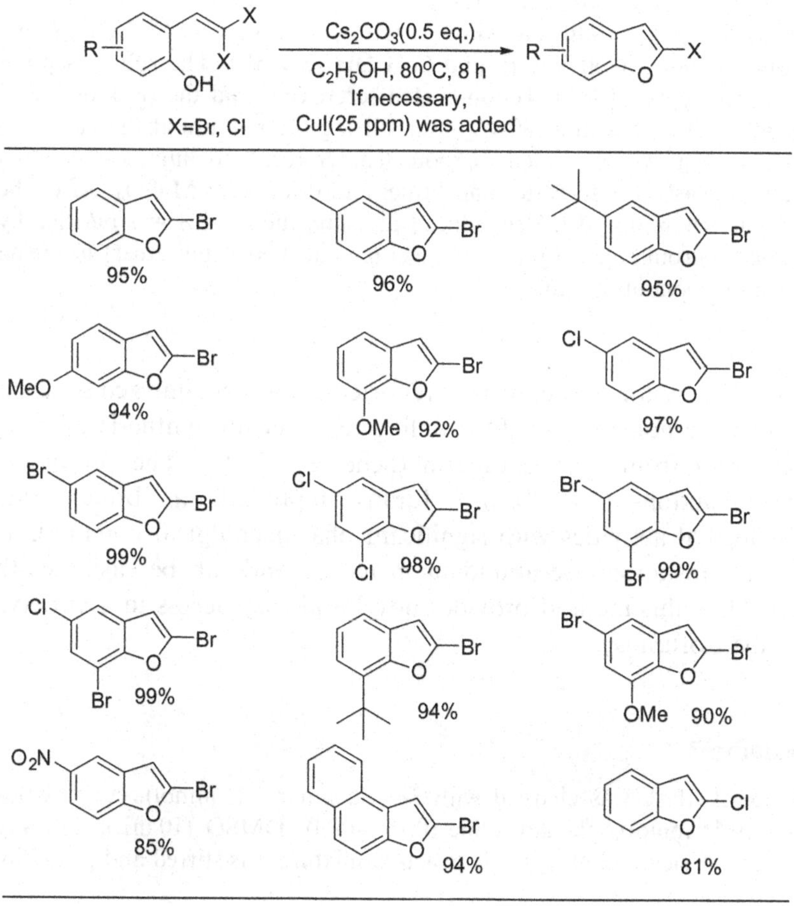

Scheme 3.6 Trace amount Cu (ppm)-catalyzed intramolecular cyclization for synthesis of 2-bromobenzofurans.

trace amount of Cu (0.0064 mol%, 25 ppm). The reaction provides the desired products in excellent yields under fluoride-free and mild reaction conditions and with a TON (turnover number) of up to 1.5×10^4. They also found that the 47 ppm of Cu in the commercially available Cs_2CO_3 (99.9% from Shanghai, China) was capable of catalyzing the intramolecular cyclization of substrates. The reactions were also completed using Cs_2CO_3 (99.995% from Aldrich) with additional CuI (25 ppm), providing comparable yields of products. This method can also be used in the synthesis of 2-chlorobenzofurans and is a mild and environmentally-friendly reaction.

Scheme 3.6

A sealable reaction tube equipped with a magnetic stirrer bar was charged with gem-dibromovinyl substrate (1.0 mmol), Cs_2CO_3 (99.9% from Shanghai, 0.50 mmol), and C_2H_5OH (2.0 mL). The rubber septum was then replaced by a Teflon-coated screw cap, and the reaction vessel placed in an oil bath at 80°C. After stirring the mixture at this temperature for 8 h, it was cooled to room temperature and diluted with ethyl acetate, washed with water and brine, and dried over $MgSO_4$. After the solvent was removed under reduced pressure, the residue was purified by column chromatography on silica gel (eluant: petroleum ether) to create the corresponding product.

In 2013, Wang and coworker reported a copper-catalyzed decarboxylative intramolecular C—O coupling reaction for synthesis of 2-arylbenzofuran from 3-arylcoumarin (Scheme 3.7) [7]. The products of these reactions—2-arylbenzo furans—represent a broad range of biological activities with significant pharmacological potential. The starting substances are abundant in nature and can be easily synthesized. Thus this method provides novel and easy access to a variety of 2-arylbenzofurans.

Scheme 3.7

A 25 mL flask was charged with 3-arylcoumarin (1 mmol), cupric chloride (0.15 mmol), phenanthroline (0.15 mmol), DMSO (10 mL), and 4 Å molecular sieves (300 mg). The reaction mixture was stirred and primarily

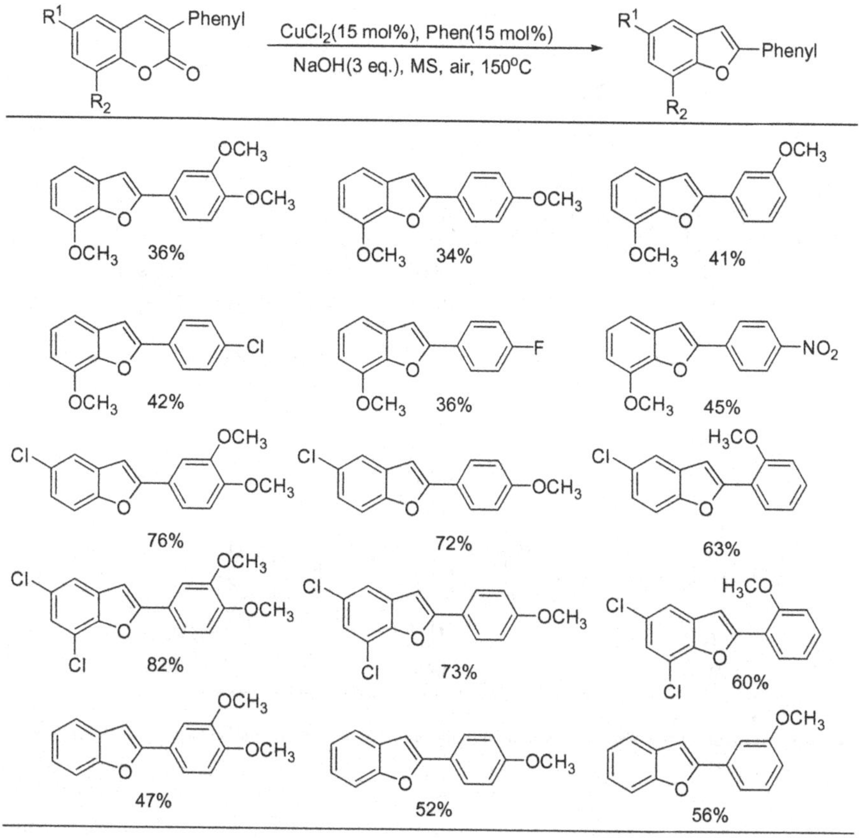

Scheme 3.7 Synthesis of 2-arylbenzofuran from 3-arylcoumarin.

heated to 110°C for 1 h and the color gradually turned to dark brown. The temperature was then raised to 150°C and maintained for 24 h. The mixture was exposed to air during the reaction time. After cooling to room temperature, hydrochloric acid (2 mol/L, 10 mL) and water (20 mL) were added to terminate the reaction, which simultaneously brought about the generation of brown solid and bubble. The suspension was then extracted with chloroform (20 mL * 3). The combined organic layer was washed in turn with water (20 mL) and then brine (20 mL), dried over anhydrous magnesium sulfate, filtered, and concentrated under reduced pressure. The solid residue obtained was purified by silica gel column chromatography.

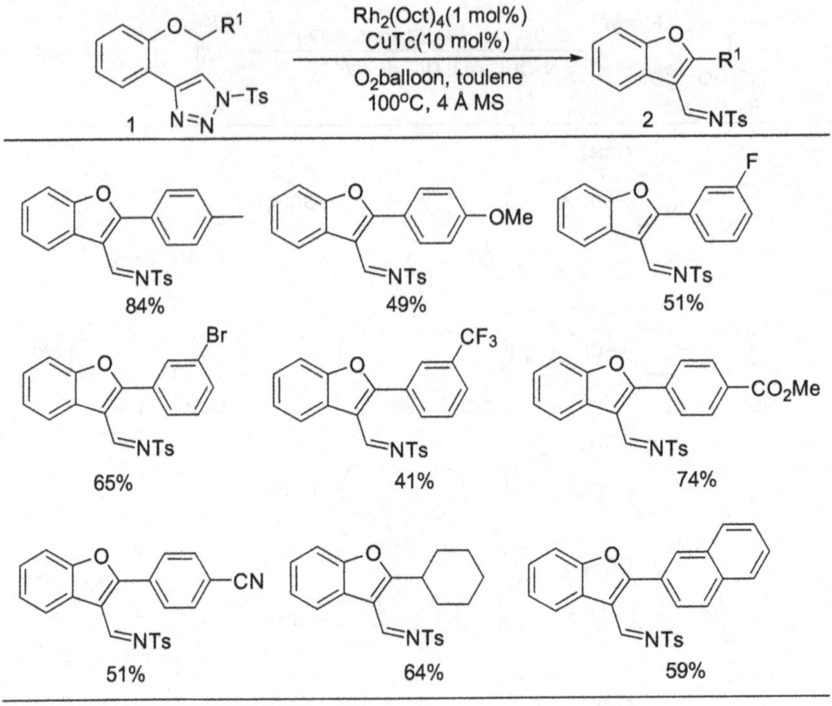

Scheme 3.8 Synthesis of benzofurans via transition metal-catalyzed tandem intramolecular $C(sp^3)-H$ insertion.

In 2015, Kang and coworker developed a synthesis of benzofurans via the transition metal-catalyzed tandem intramolecular $C(sp^3)-H$ insertion of azavinyl carbenes, derived from 1-sulfonyl-1,2,3-triazoles and aerobic oxidation via sequential catalysis (Scheme 3.8) [8]. The reaction has good tolerance to different groups. Substrates with electron withdrawing or electron-donating substituents on different positions of the phenyl ring are also converted to the desired products with moderate to good yields.

Scheme 3.8

Triazole derivative 1 (0.3 mmol), 4 Å MS (20 mg), CuTc (0.03 mmol, 0.1 eq.), and $Rh_2(Oct)_4$ (0.003 mmol, 0.01 eq.) were added to an oven-dried Schlenk tube equipped with a stir bar. The reaction vessel was evacuated and backfilled with O_2 three times before adding freshly distilled toluene (6.0 mL, 0.05 M). The reaction mixture was stirred under an O_2 (balloon) atmosphere at 100°C (checked by thin layer chromatography (TLC)). The residue was purified by flash column chromatography with ethyl acetate and petroleum ether as eluents to afford 2.

In 2006, Ackermann and coworker also developed copper-catalyzed benzo[b]furan synthesis (Scheme 3.9) [9]. Aryl chlorides can also be used in this reaction.

In 2011, Wang and coworker developed a copper-catalyzed coupling reaction for synthesis of benzofurans and indoles (Scheme 3.10) [10]. The reaction used CuBr as the catalyst, Cs_2CO_3 as the base, and MeCN as the solvent. The reaction also has a wide scope of starting substances. Substrates with electron-withdrawing or electron-donating substituents on different positions of the phenyl ring can be tolerated and converted to the desired products with moderate to good yields.

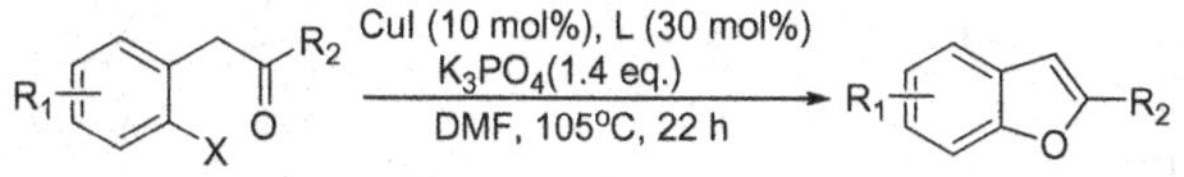

Entry	Starting substance	L	Product	Yield (%)
1	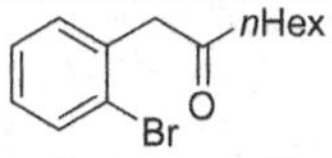	–	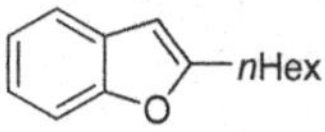	82
2	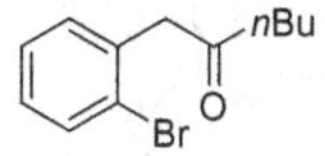	–	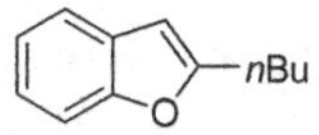	82
3	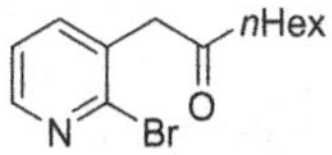	–	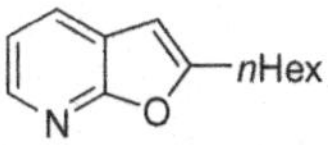	80
4	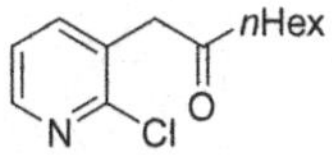	–	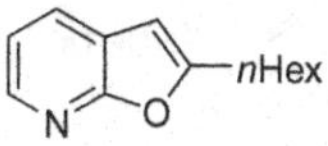	56
5	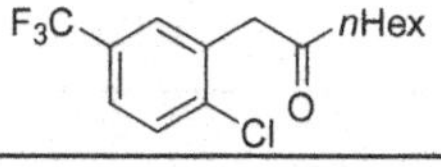	$Me_2NCH_2CO_2H$	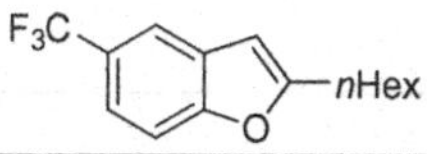	69

Scheme 3.9 Copper-catalyzed benzofuran synthesis.

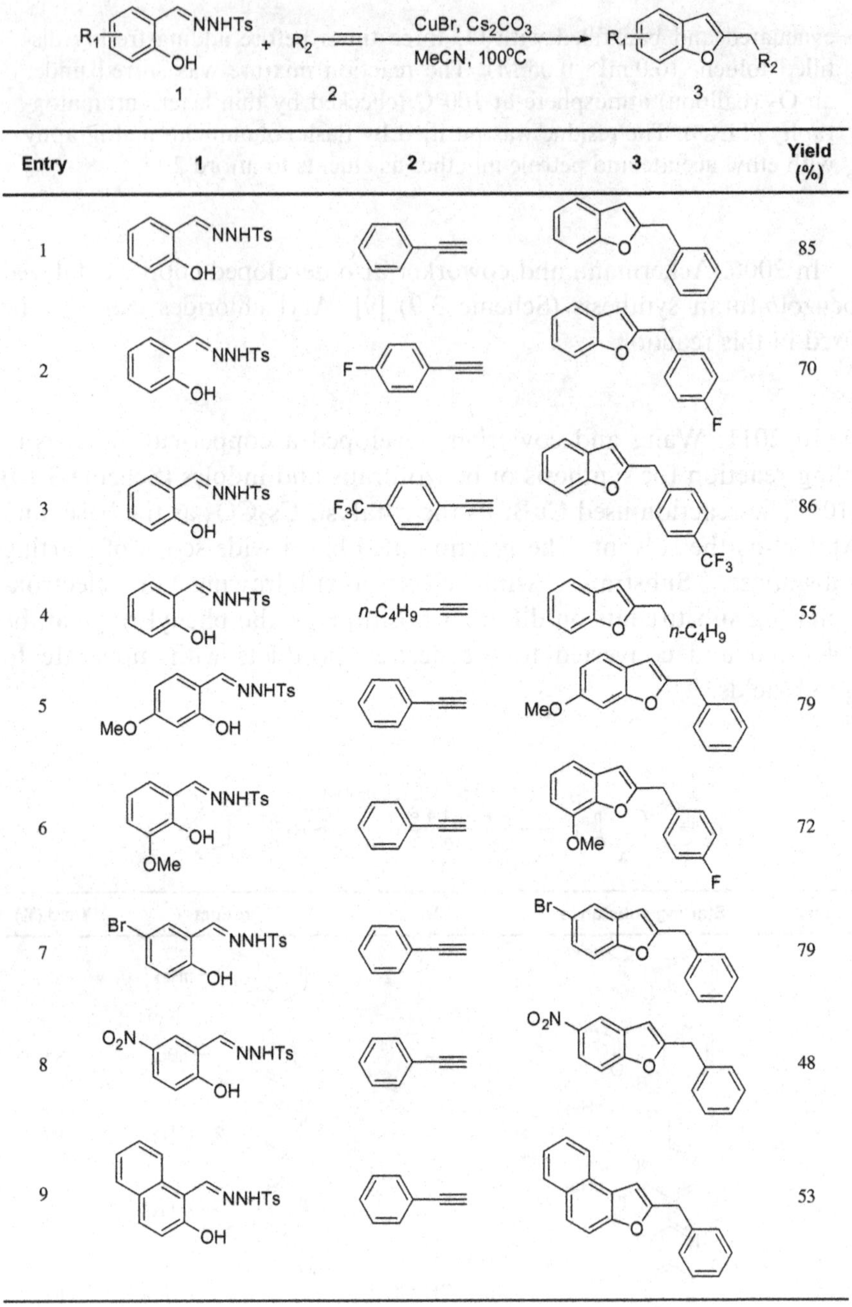

Entry	1	2	3	Yield (%)
1				85
2				70
3				86
4				55
5				79
6				72
7				79
8				48
9				53

Scheme 3.10 Cu-catalyzed reaction of N-tosylhydrazone and various terminal alkynes.

REFERENCES

[1] Chen, C.-Y.; Dormer, P. G. *J. Org. Chem.* **2005,** *70*, 6964–6967.

[2] Carril, M.; SanMartin, R.; Tellitu, I.; Domínguez, E. *Org. Lett.* **2006,** *8*, 1467–1470.

[3] Nagamochi, M.; Fang, Y.-Q.; Lautens, M. *Org. Lett.* **2007,** *9*, 2955–2958.

[4] Newman, S. G.; Aureggi, V.; Bryan, C. S.; Lautens, M. *Chem. Commun.* **2009,** 5236–5238.

[5] Zhou, W.; Chen, W.; Wang, L. *Org. Biomol. Chem.* **2012,** *10*, 4172–4178.

[6] Ji, Y.; Li, P.; Zhang, X.; Wang, L. *Org. Biomol. Chem.* **2013,** *11*, 4095–4101.

[7] Pu, W.-C.; Mu, G.-M.; Zhang, G.-L.; Wang, C. *RSC Adv* **2014,** *4*, 903–906.

[8] Li, L.; Xia, X.-H.; Wang, Y.; Bora, P. P.; Kang, Q. *Adv. Synth. Catal.* **2015,** *357*, 2089–2097.

[9] Ackermann, L.; Kaspar, L. T. *J. Org. Chem.* **2007,** *72*, 6149–6153.

[10] Zhou, L.; Shi, Y.; Xiao, Q.; Liu, Y.; Ye, F.; Zhang, Y., et al. *Org. Lett.* **2011,** *13*, 968–971.

CHAPTER 4

Other Transition Metal-Catalyzed Benzofuran Synthesis

In 2004 Wang and coworker reported a versatile and new method for the synthesis of benzofurans from various phenols by the following strategy: (1) various allyl phenyl ethers (2a-f) prepared from O-alkylation of various phenols (1a-f) with corresponding alkyl halide underwent [3,3] sigmatropic (Claisen) rearrangement to furnish O-allylphenols (3a-f), respectively; (2) then, (3a-f) underwent O-chloroethylation with an excess of 1,2-dichloroethane, sodium hydroxide in water, and terabutylammonium bromide as phase catalyst to give monoalkylated products (4a-f); (3) treatment of compound 4a-f with potassium tertbutoxide in THF underwent isomerization of the allyl group together with 1,2-elimination of O-(2-chloroethyl) group to build up the O-vinyl and C-propenyl function as precursor (5a-f) for RCM; (4) finally, the cyclization of compound 5a-f with Grubbs' catalyst underwent RCM to create various benzofurans (6a-f) with good overall yields (Scheme 4.1) [1].

In 1998 Furukawa and coworker developed Ru and Cu cocatalyzed cyclization of 2-allylphenol to 2,3-dihydro-2-methylbenzofuran without β-elimination (Scheme 4.2) [2]. The intramolecular nuclephilic addition of 2-allylphenol was catalyzed by $RuCl_3$/AgOTf-PPh$_3$-Cu(OTf)$_2$ to afford 2,3-dihydro-2-methylbenzofuran with good yields.

In 2009 Hashmi and coworker reported a gold-catalyzed reaction for synthesis of benzo[b]furans. The mono-substituted furans could be converted to benzo[b]furans with low yields (Scheme 4.3) [3]. The reactions are unselective, but the products could easily be separated from the unknown byproducts of higher polarity, which are likely oligomers/polymers. With one methyl group as a donor substituent in the 5-position of the furan, which increases the nucleophilicity of the furan ring, much higher yields could be obtained.

In 2013 Pyne and coworker developed concise synthesis of α-substituted 2-benzofuranmethamines and 2-subsituted benzofurans via α-substituted

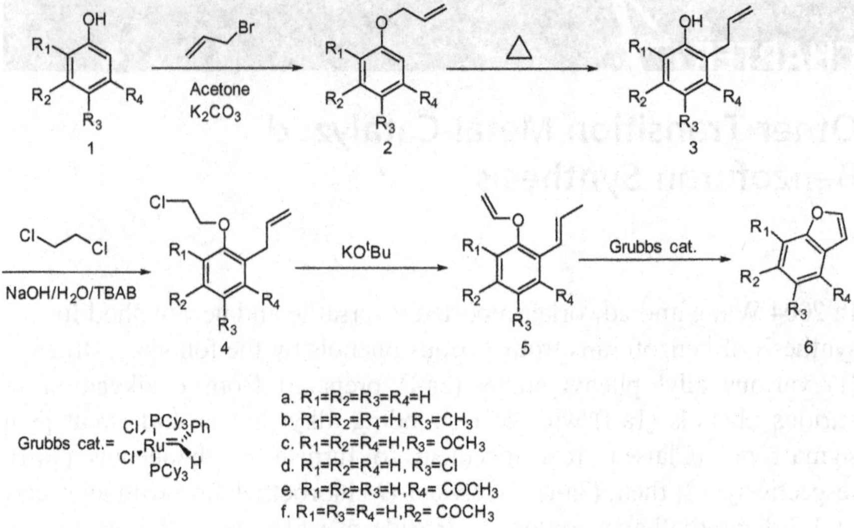

Scheme 4.1 New synthesis of benzofurans from phenols via claisen rearrangement and ring-closing metathesis.

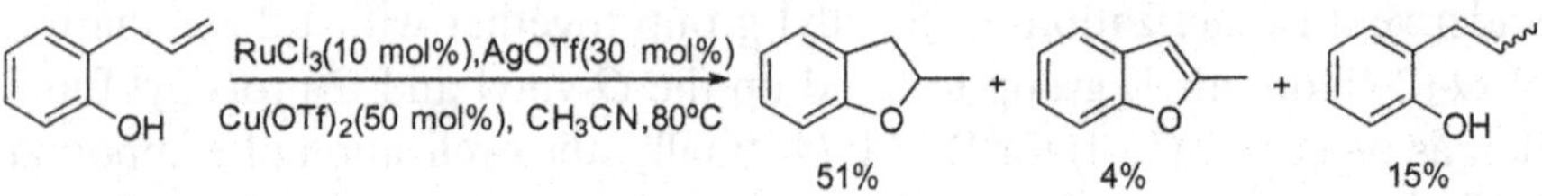

Scheme 4.2 New synthesis of benzofurans from phenols.

2-benzo-furanmethyl carbocation intermediates (Scheme 4.4) [4]. The reaction used $AgNO_3$ as the catalyst and reacted in hot DMF. α-Substituted 2-benzofuran-methamine derivatives could be obtained with moderate to good yields through sequential cycloisomerization-1,3-allylic rearrangement.

In 2014 Bi and coworker developed new silver-catalyzed heteroaromatization of propargylic alcohols with p-toluenesulfonylmethyl isocyanide, which provides an efficient and modular approach to sulfonyl benzoheteroles (Scheme 4.5) [5]. In this case, sulfonylation and cyclization seemed to occur, leading to α-sulfonyl ketones as intermediates. Upon cyclization and aromatization of the latter, sulfonyl benzofurans were obtained with good overall yields.

In 2004 Fan and coworker developed silver-catalyzed synthesis of 4-substituted benzofurans via a cascade oxidative coupling-annulation protocol (Scheme 4.6) [6]. This one-pot cascade reaction involved

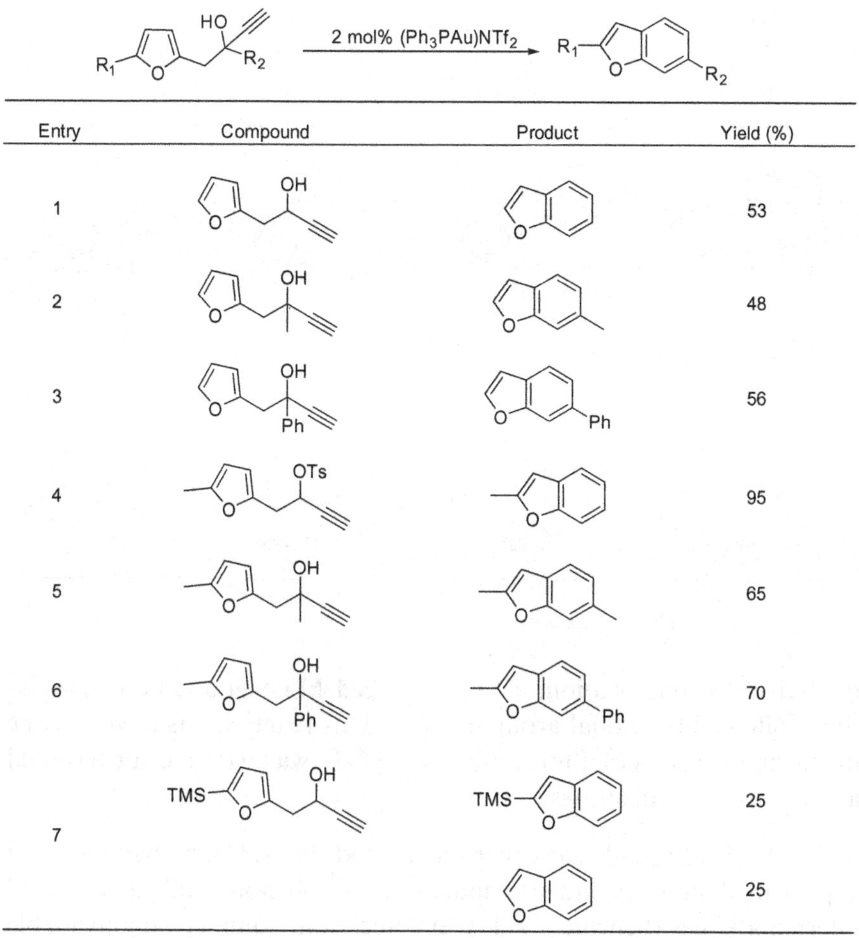

Entry	Compound	Product	Yield (%)
1			53
2			48
3			56
4			95
5			65
6			70
7			25
			25

Scheme 4.3 Gold-catalyzed synthesis of benzofurans.

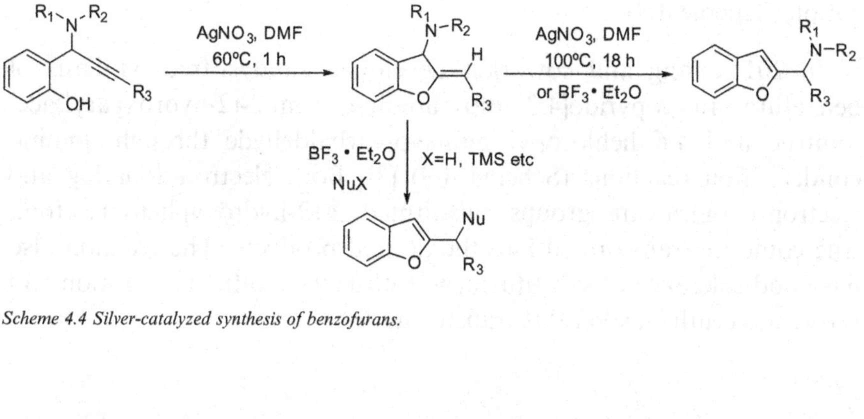

Scheme 4.4 Silver-catalyzed synthesis of benzofurans.

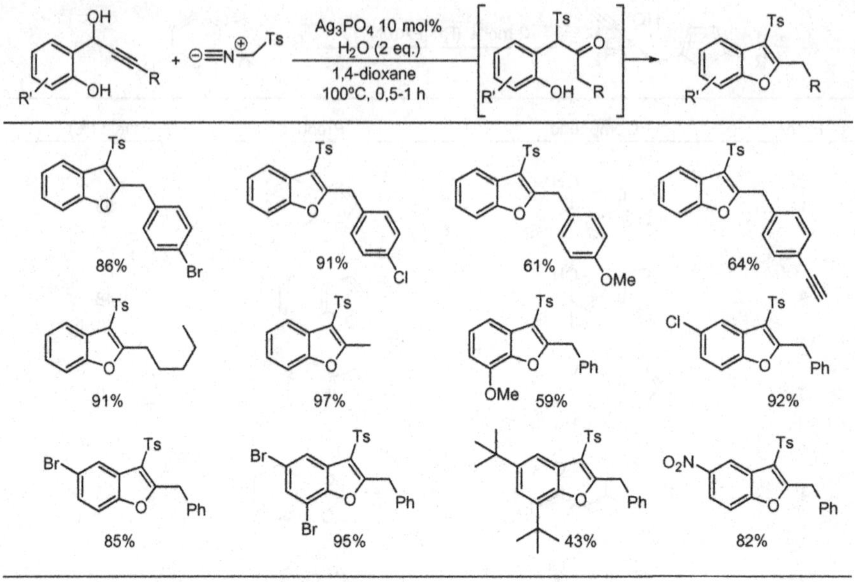

Scheme 4.5 Synthesis of 3-sulfonyl benzofurans.

oxidative dearomatization, a silver-induced Michael addition—annulation, followed by a final aromatization. This reaction has a wide scope in starting substances. Stoichiometric AgOTf was required for terminal alkynes used as substrates.

In 2013 Shi and coworker developed benzofuran synthesis via copper-mediated oxidative annulation of phenols and unactivated internal alkynes (Scheme 4.7) [7]. Starting from commercially available phenols and alkynes, direct one-step/pot synthesis of benzofuran derivatives can be achieved. The researchers also studied the mechanism in which annulations of alkynes with phenols through reversible electrophilic carbocupration of phenol followed by alkyne insertion and cyclization (Scheme 4.8).

In 2014 Yang and coworker developed catalyst-free synthesis of benzofuran-fused pyrido[4,3−d]pyrimidines from 2-(2-hydroxyaryl)acetonitrile and 4,6-dichloropyrimidine-5-carbaldehyde through domino condensation reactions (Scheme 4.9) [8]. Both electron-donating and electron-withdrawing groups substituted 2-(2-hydroxyphenyl)acetonitrile could be transformed into the desired products. The reaction also has good tolerance of substituent. Substituent on different position had no signiant influence on this transformation.

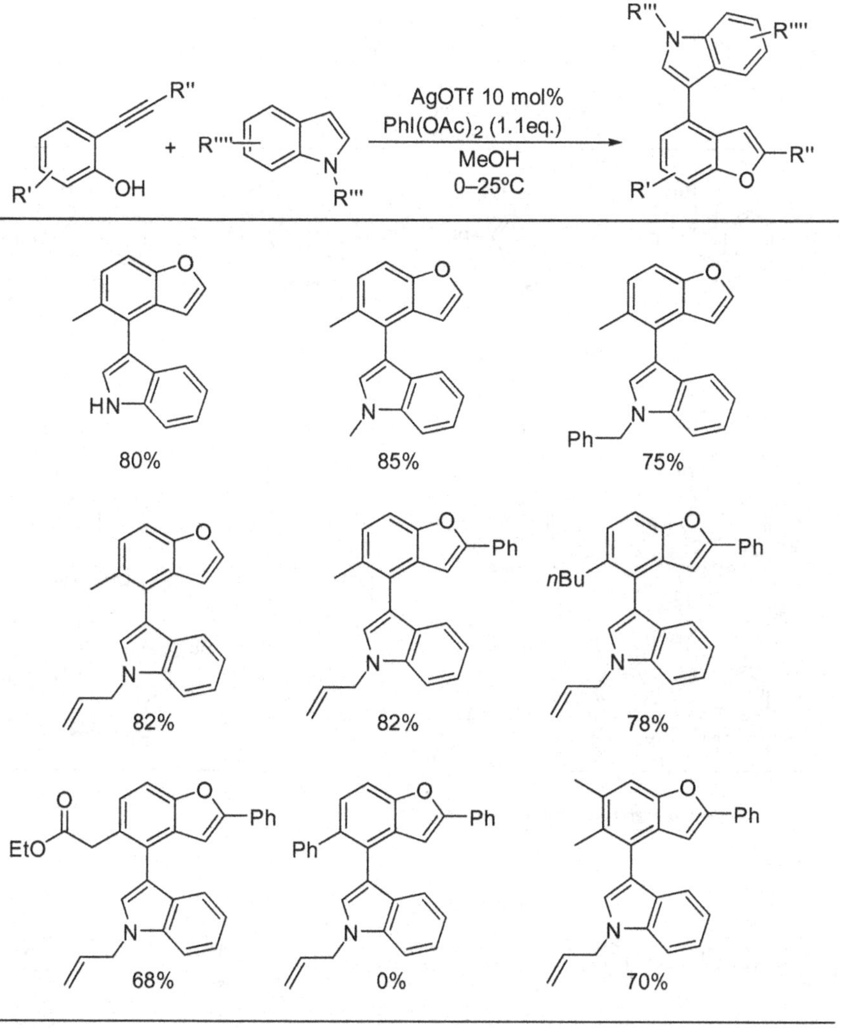

Scheme 4.6 Silver-catalyzed synthesis of 4-substituted benzofurans.

The researchers also reported a possible mechanism. The starting materials underwent nucleophilic aromatic substitution (S_NAr) reaction to obtain intermediate A, followed by sequential cyclization to form the intermediate B, which could undergo isomerization to afford C. Finally, the intermediate C underwent dehydrated aromatization to form the product 3a (Scheme 4.10).

In 1995 Dufiach and coworker developed electrochemical intramolecular reductive cyclization catalyzed by electrogenerated Ni

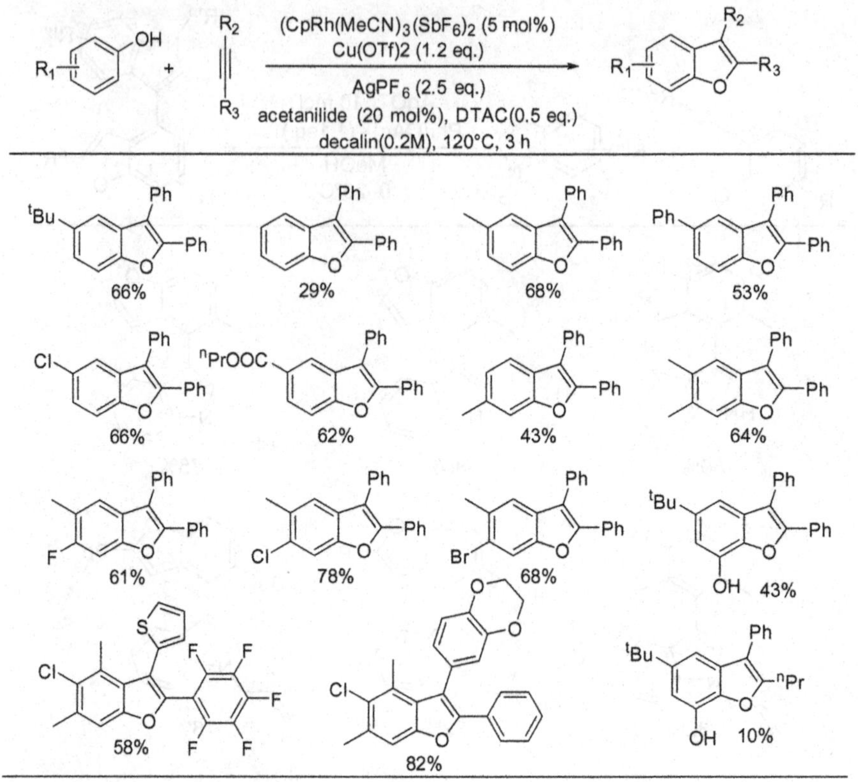

Scheme 4.7 Synthesis of highly substituted benzofurans.

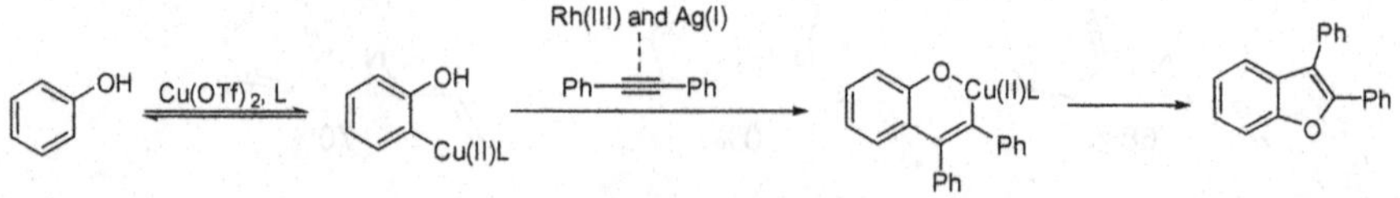

Scheme 4.8 Proposed mechanism.

$(cyclam)^{2+}$ (Scheme 4.11) [9]. One of the features of this reaction is a series of *O*-halogenated aromatic compounds containing unsaturateds ide-chains that were used and resulted in good to excellent yields.

In 2015 Sun and coworker reported zeolite-catalyzed synthesis of substituted benzo[b]furans (Scheme 4.12) [10]. The reaction was carried out through the intramolecular cyclization of 2-aryloxyacetaldehyde acetals. Interestingly, zeolite was used as the catalyst, which was always used in organic synthesis as one of solid acids. However, the

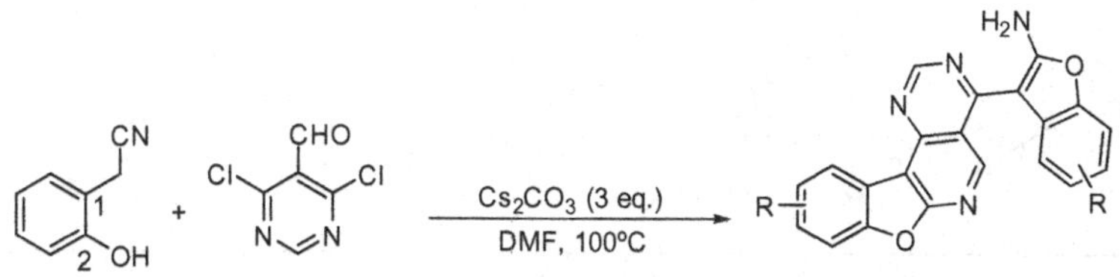

Entry	R	Yield (%)
1	H	84
2	3-OCH₃	68
3	3-F	64
4	4-Br	73
5	4-N(CH₃)₂	58
6	4-OCH₃	51
7	4-CF₃	57
8	5-F	53

Scheme 4.9 Synthesis of 2-aminobenzofuran.

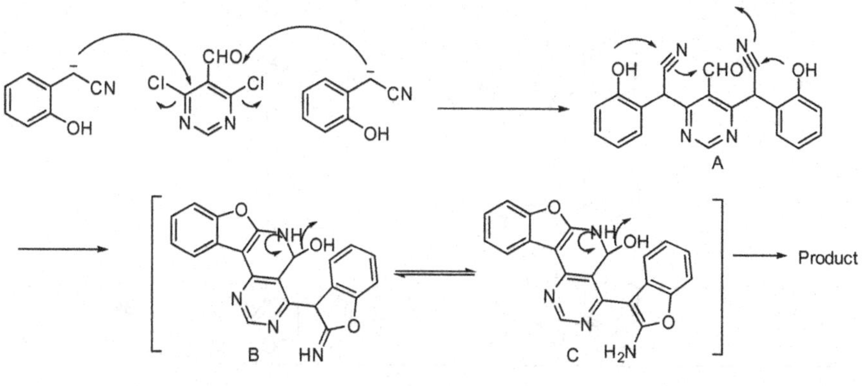

Scheme 4.10 Proposed mechanism.

reaction can only been used in the synthesis of nonsubstituted benzo[b] furans. The position of the substituent on the phenyl ring also has an effect on the reaction.

In 2013 Nolan and coworker developed gold-catalyzed decarboxylation of aromatic carboxylic acid, which can be used in the synthesis of benzofuran (Scheme 4.13) [11]. In the presence of (Au(SIPr)O₂CAd)

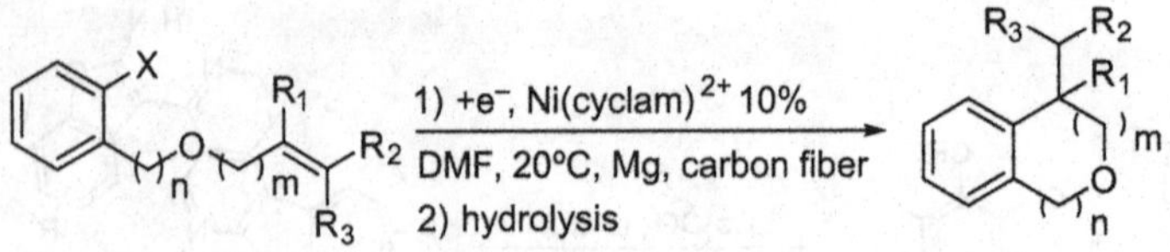

Entry	Strating substrate	Product	Yield (%)
1	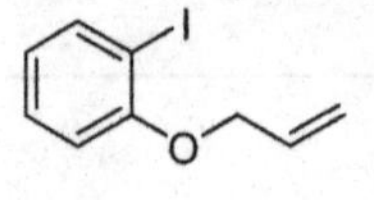	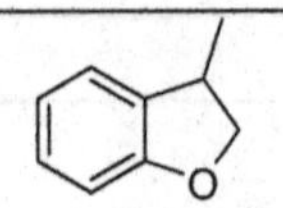	90
2	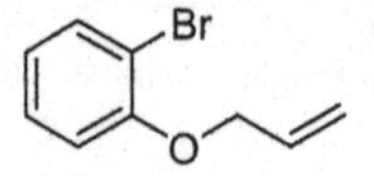	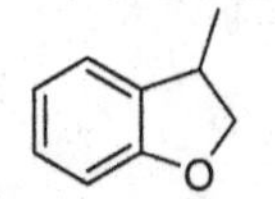	86
3	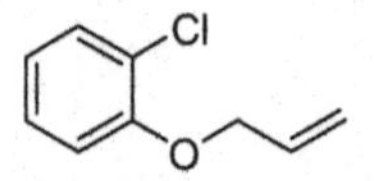	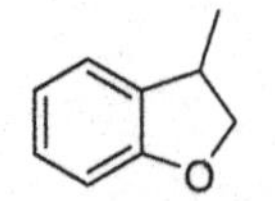	60
4	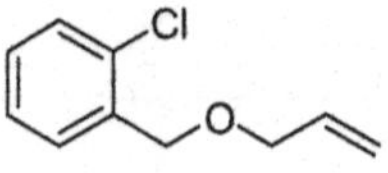	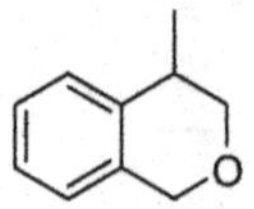	56
5	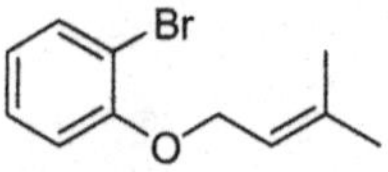	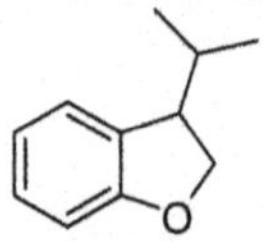	62
6	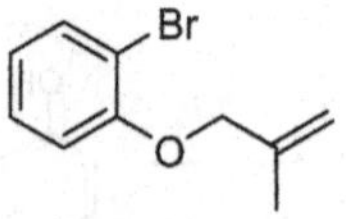	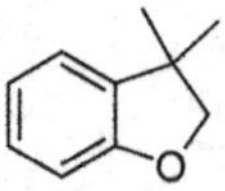	54
7	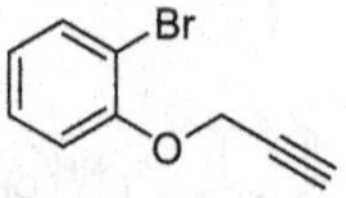	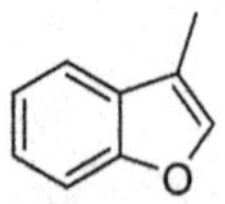	32
8	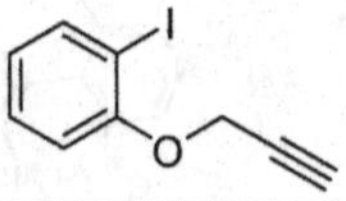	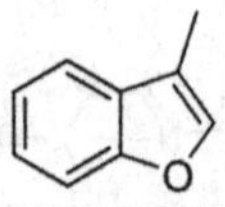	31

Scheme 4.11 Electrochemical intramolecular cyclization.

and 1 eq. of 1-Adamantanecarboxylic acid (AdCOOH), the reaction can be processed. Interestingly, *ortho*-substituted substrates and pentafluorobenzoic acid also react well.

In 2006 Sebastiani and coworker reported a TiO_2-sensitized photooxidation reaction of indane and some of its hetero-analogs in deaerated CH_3CN and in the presence of Ag_2SO_4 (Scheme 4.14) [12].

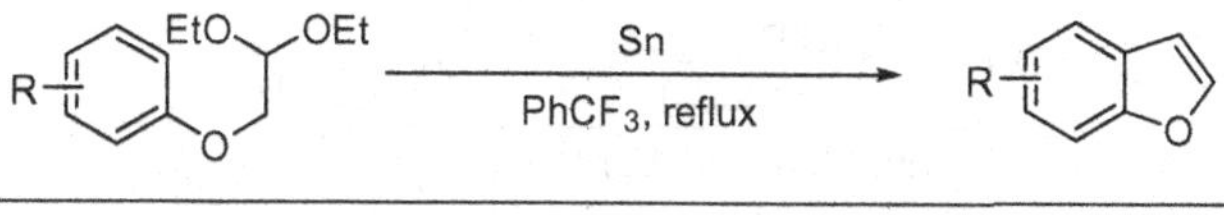

Entry	Strating substrate	Product	Yield (%)
1	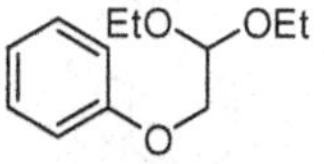	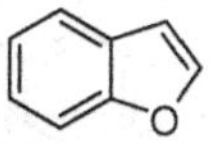	78
2	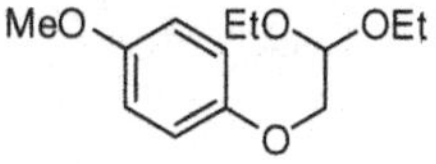	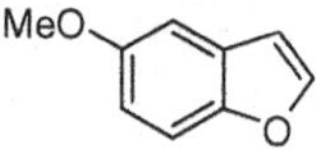	86
3	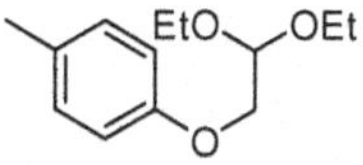	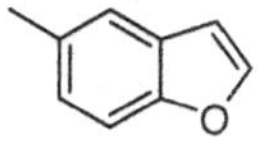	93
4	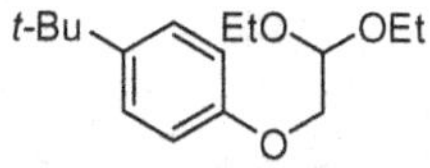	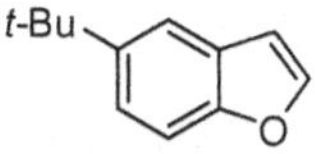	92
5	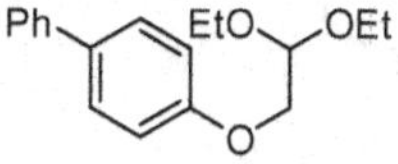	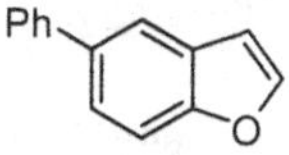	95
6	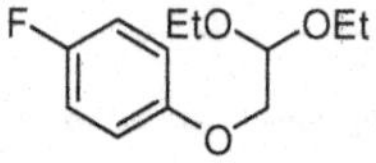	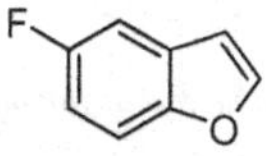	67
7	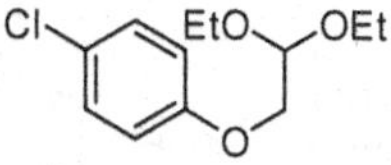	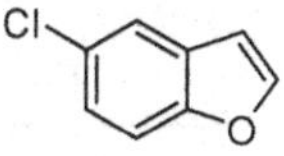	80
8	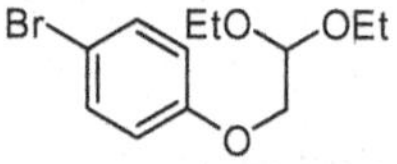	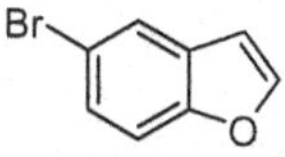	77
9	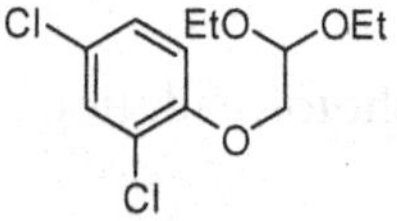	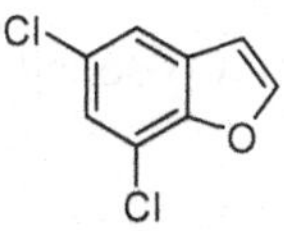	30
10	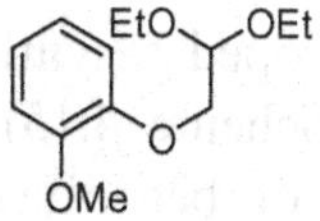	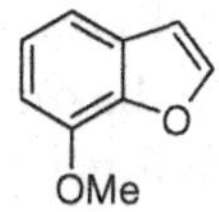	90
11	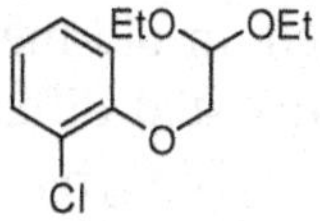	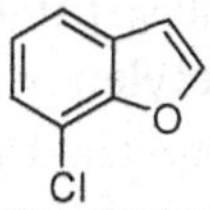	75

Scheme 4.12 Preparation of various 2, 3-unsubstituted benzo[b]furans.

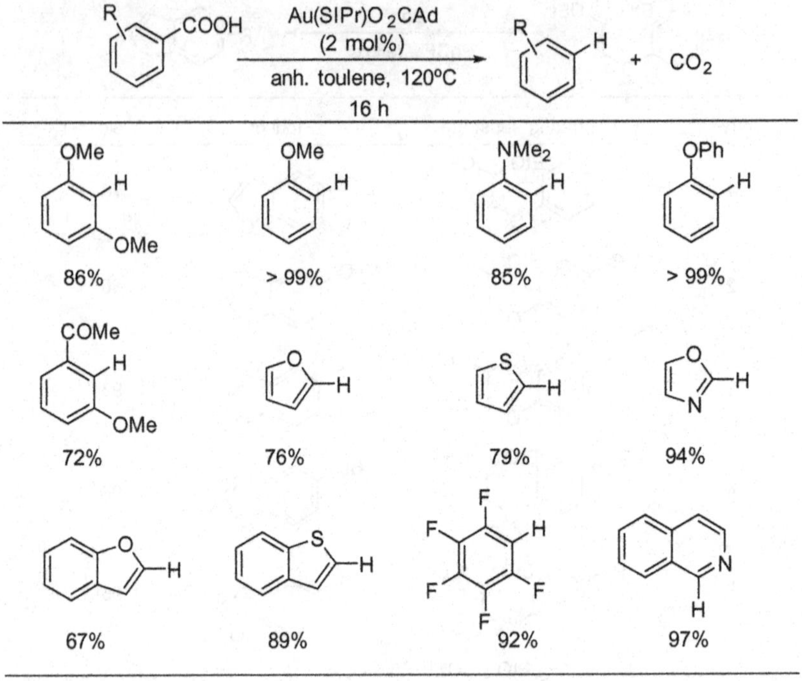

Scheme 4.13 Gold-catalyzed protodecarboxylation of (hetero)aromatic carboxylic acids.

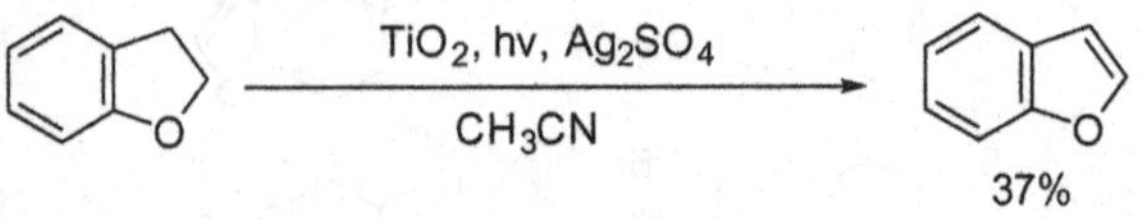

Scheme 4.14 TiO₂-sensitized photo-oxidation of 2,3-dihydrobenzofuran.

They also studied TiO_2-sensitized photo-oxidation of indane and its hetero-analogs.

In 1997 Shioiri and coworker developed a synthesis of benzofuran utilizing trimethyl-silydiazomethane (Scheme 4.15) [13]. The method contained two steps for preparation of benzofuran from siloxyaryl ketones and aldehydes. First, they used an old method for preparation of o-siloxyphenylacetylenes. They had tested different o-trisiloxyaceto-phenones. Then, they used o-triisopropylsioxyphenylacetylenes to pre-pare the benzofuran.

Otterlo and coworker developed a new strategy for the synthesis of substituted benzofuran (Scheme 4.16) [14]. The reaction has a wide

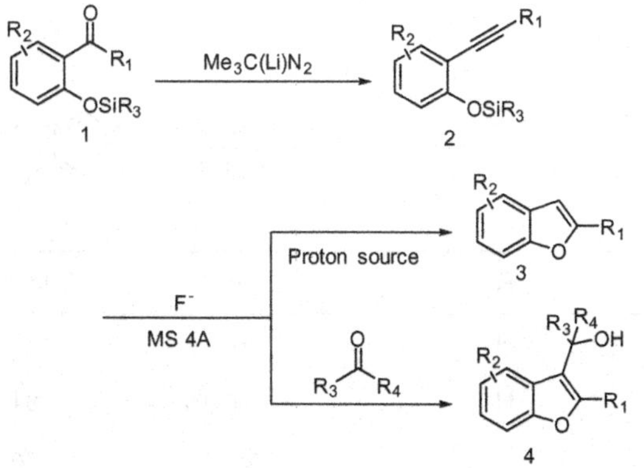

	Preparation of 2		
Entry	R_1	R_2	Yield (%)
1	Me	H	78
2	Et	H	65
3	H	H	69
4	H	5-Cl	55
5	H	3-MeO	74

	Preparation of Benzofuran 3		
Entry	R_1	R_2	Yield (%)
1	Me	H	71
2	Et	H	75
3	H	H	56
4	H	5-Cl	76
5	H	3-MeO	75

	Preparation of 3-benzofuranmethanols 4				
Entry	R_1	R_2	R_3	R_4	Yield (%)
1	Me	H	Ph	H	81
2	Et	H	Ph	H	80
3	H	H	Ph	H	67
4	H	5-Cl	Ph	H	77
5	H	3-MeO	Ph	H	86
6	Me	H	4-MeOC$_6$H$_4$	H	65
7	Me	H	4-ClC$_6$H$_4$	H	81
8	Me	H	2-Furyl	H	76
9	Me	H	2-Thienyl	H	70

Scheme 4.15 Synthesis of benzofuran.

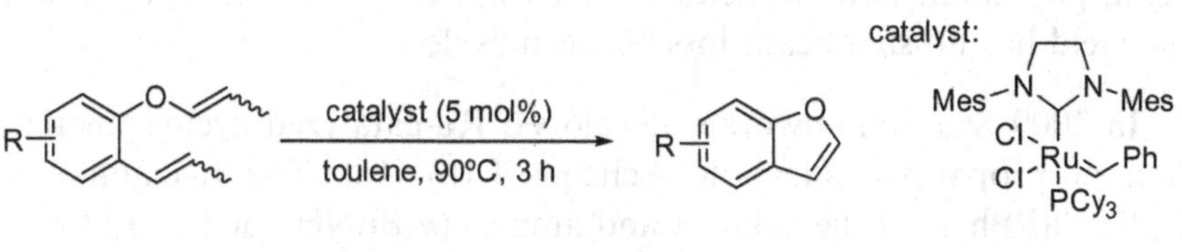

Scheme 4.16 Synthesis of benzofuran with Ruthenium as the catalyst.

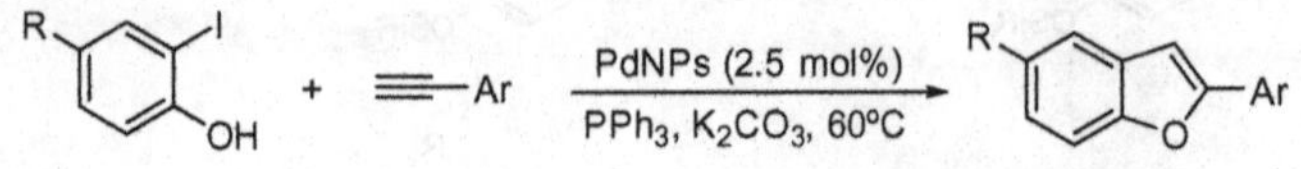

Entry	R	Ar	Yield[b] (%)
1	H	Ph	79
2	H	2-MeC$_6$H$_4$	81
3	H	3-MeC$_6$H$_4$	70
4	H	4-MeC$_6$H$_4$	72
5	H	4-t-BuC$_6$H$_4$	78
6	H	3-H$_2$NC$_6$H$_4$	86
7	H	4-H$_2$NC$_6$H$_4$	81
8	H	3-pyridyl	81
9	H	3-HC≡CC$_6$H$_4$	68
10	H	3-F$_3$CC$_6$H$_4$	75
11	H	4-NCC$_6$H$_4$	81
12	H	2,4-F$_2$C$_6$H$_3$	78
13	H	4-BrC$_6$H$_4$	87
14	CO$_2$Me	Ph	86
15	CO$_2$Me	4-NCC$_6$H$_4$	87

Scheme 4.17 Synthesis of aryl-substituted benzofurans.

scope of possible substances. Both the donating group and electron-withdrawing group can be tolerated in the reaction. The yields of most reactions are also excellent.

In 2015 Chand and coworker reported palladium nanoparticles-catalyzed synthesis of benzofurans (Scheme 4.17) [15]. The feature of using palladium nanoparticles is that the catalyst can be reused, and the yield has no significant loss between cycles.

In 2009 Saá and coworker developed Ru-catalyzed cycloisomerization of propargylic alcohols (Scheme 4.18) [16]. The reaction used CpRuCl(PPh$_3$)$_2$ as the catalyst and amines (n-BuNH$_2$ or Py) and both

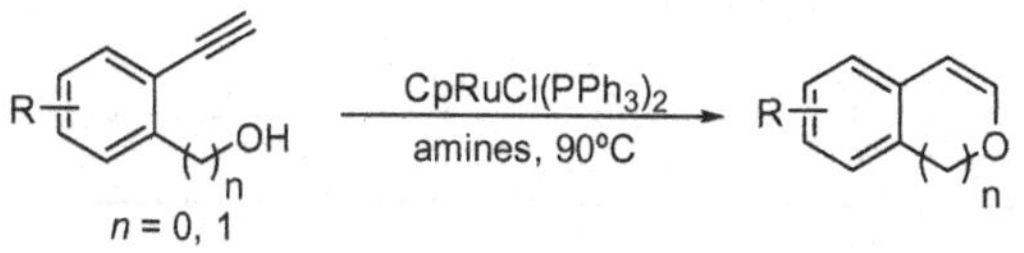

$$R \underset{}{\overset{}{\text{(arene)}}} \text{—C≡CH, —OH}, n = 0, 1 \xrightarrow[\text{amines, 90°C}]{\text{CpRuCl(PPh}_3)_2} R \text{—(product)}_n$$

Entry	Substrate	Product	Yield (%)
1	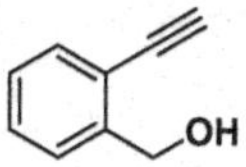	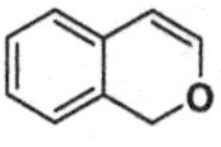	86
2	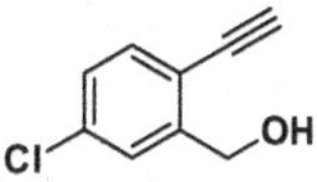 Cl	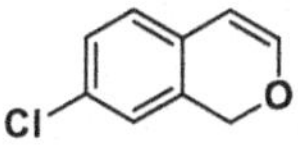 Cl	62
3	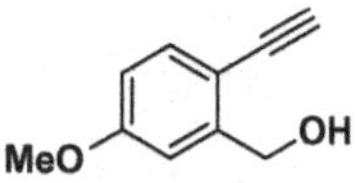 MeO	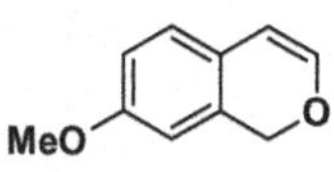 MeO	63
4	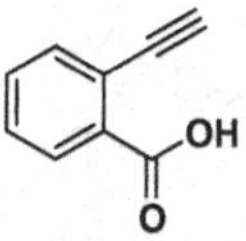	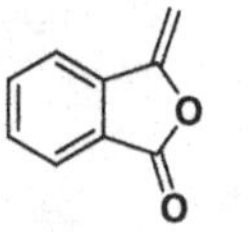	85
5	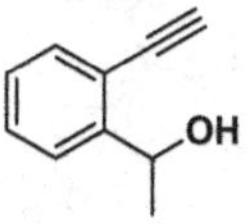	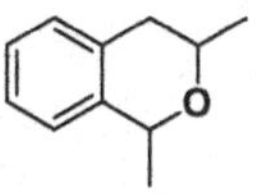	61
6	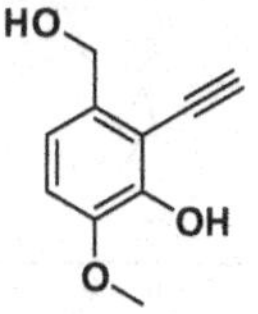 HO, O—	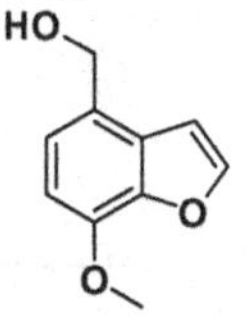 HO, O—	82
7	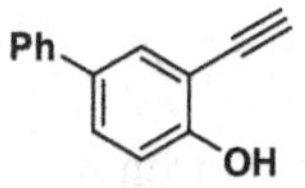 Ph	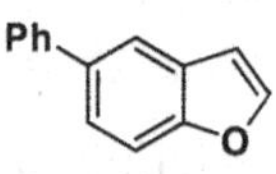 Ph	84
8	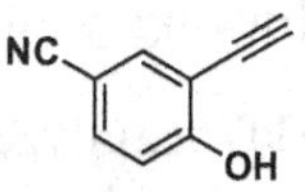 NC	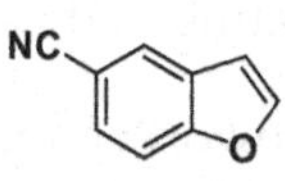 NC	54
9	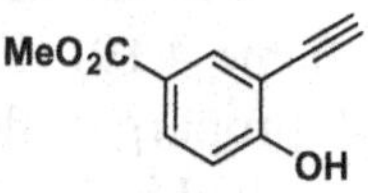 MeO$_2$C	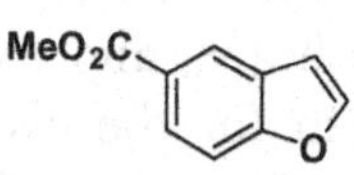 MeO$_2$C	59

Scheme 4.18 Ru-catalyzed cycloisomerization.

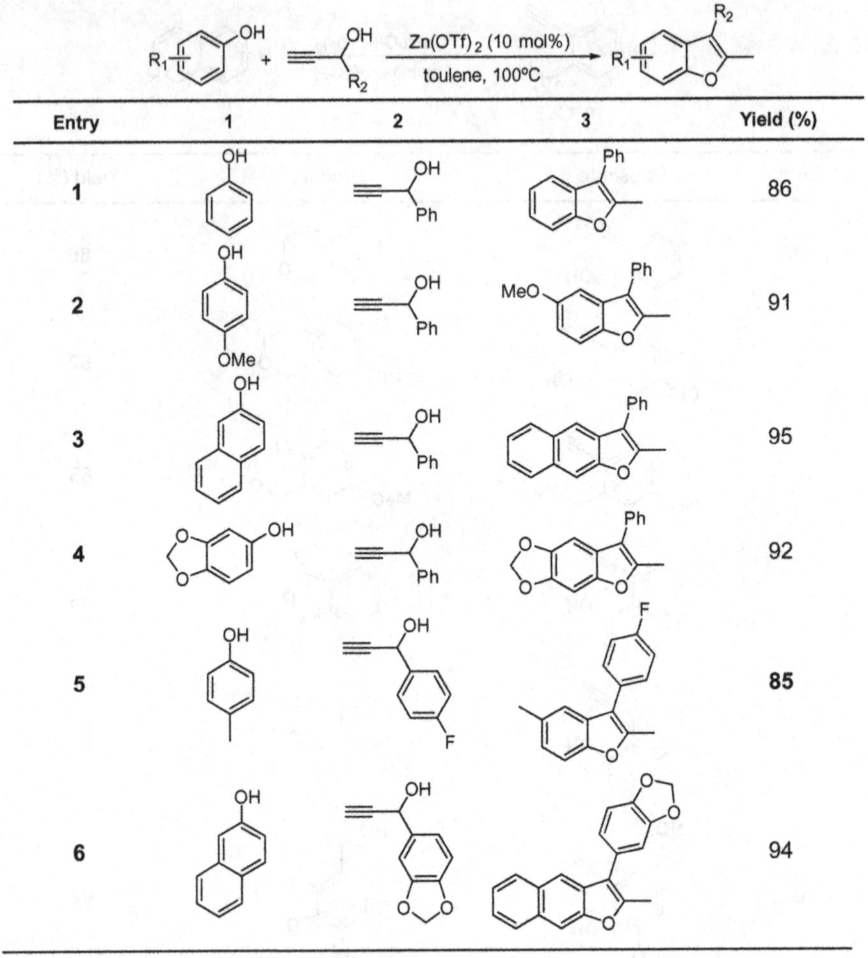

Scheme 4.19 Zn(OTf)₂-catalyzed synthesis of benzofurans.

were essential for the reaction. The reactions also have good tolerance for both electron-donating and electron-withdrawing groups.

In 2006 Liu and coworker reported $Zn(OTf)_2$-catalyzed cyclization of proparyl alcohols with anilines, phenols, and amides for the synthesis of indoles, benzofurans, and oxazoles (Scheme 4.19) [17]. 10 mol% of $Zn(OTf)_2$ was used for the cyclization of propargyl alcohols with PhOH or $PhNH_2$ in toluene (100°C), which resulted in indole and benzofuran products with excellent or good yields. However, this reaction is sensitive to the solvent (it was found that the best solvent is toluene).

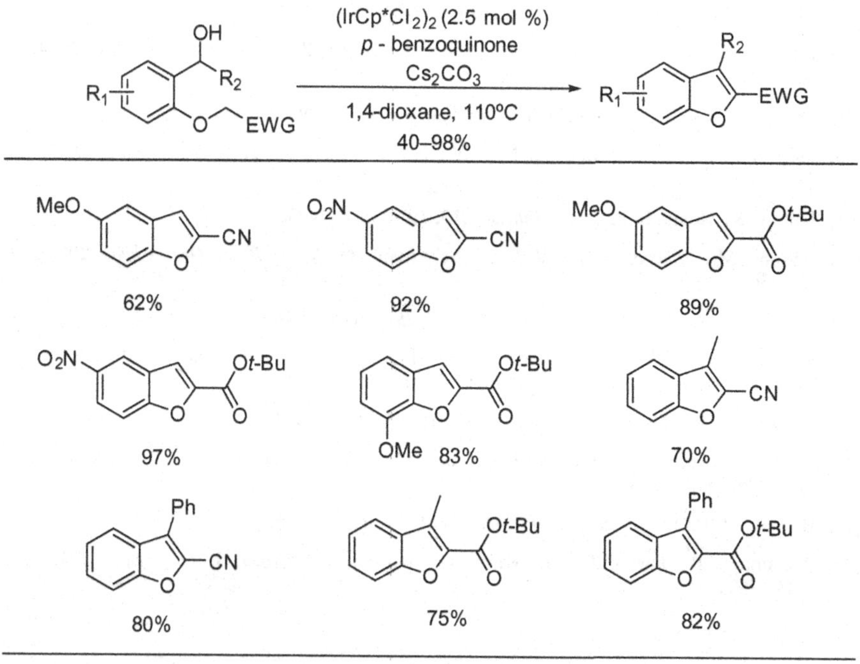

Scheme 4.20 Ir-catalyzed synthesis of benzofurans.

In 2013 Cossy and coworker reported iridium-catalyzed hydrogen transfer for synthesis of substituted benzofurans (Scheme 4.20) [18]. In this transformation, the presence of various electron-withdrawing groups was not detrimental to the process. The authors also studied substitution of the aromatic ring and found that both electron-donating and electron-withdrawing groups on aromatic ring can be tolerated.

REFERENCES

[1] Tsai, T.-W.; Wang, E.-C.; Li, S.-R.; Chen, Y.-H.; Lin, Y.-L.; Wang, Y.-F., et al. *J. Chin. Chem. Soc.* **2004**, *51*, 1307–1318.

[2] Kazushige, H.; Hideki, K.; Akio, M.; Tetsuo, O.; Isao, F. *Chem. Lett.* **1998**, *27*, 1083–1084.

[3] Hashmi, A. S. K.; Wölfle, M. *Tetrahedron* **2009**, *65*, 9021–9029.

[4] Wongsa, N.; Sommart, U.; Ritthiwigrom, T.; Yazici, A.; Kanokmedhakul, S.; Kanokmedhakul, K., et al. *J. Org. Chem.* **2013**, *78*, 1138–1148.

[5] Liu, J.; Liu, Z.; Liao, P.; Bi, X. *Org. Lett.* **2014**, *16*, 6204–6207.

[6] Ye, Y.; Fan, R. *Chem. Commun.* **2011**, *47*, 5626–5628.

[7] Zhu, R.; Wei, J.; Shi, Z. *Chem. Sci.* **2013**, *4*, 3706−3711.

[8] Li, B.; Yue, Z.; Xiang, H.; Lv, L.; Song, S.; Miao, Z.; Yang, C. *RSC Adv.* **2014**, *4*, 358−364.

[9] Olivero, S.; Clinet, J. C.; Duñach, E. *Tetrahedron Lett.* **1995**, *36*, 4429−4432.

[10] Sun, N.; Huang, P.; Wang, Y.; Mo, W.; Hu, B.; Shen, Z., et al. *Tetrahedron* **2015**, *71*, 4835−4841.

[11] Dupuy, S.; Nolan, S. P. *Chem. Eur. J* **2013**, *19*, 14034−14038.

[12] Bettoni, M.; Giacco, T. D.; Rol, C.; Sebastiani, G. V. *J. Phys. Org. Chem.* **2006**, *19*, 359−364.

[13] Ito, Y.; Aoyama, T.; Shioiri, T. *Synlett* **1997**, *1997*, 1163−1164.

[14] van Otterlo, W. A. L.; Morgans, G. L.; Madeley, L. G.; Kuzvidza, S.; Moleele, S. S.; Thornton, N., et al. *Tetrahedron* **2005**, *61*, 7746−7755.

[15] Mandali, P. K.; Chand, D. K. *Synthesis* **2015**, *47*, 1661−1668.

[16] Varela-Fernández, A.; González-Rodríguez, C.; Varela, J. A.; Castedo, L.; Saá, C. *Org. Lett.* **2009**, *11*, 5350−5353.

[17] Kumar, M. P.; Liu, R.-S. *J. Org. Chem.* **2006**, *71*, 4951−4955.

[18] Anxionnat, B.; Pardo, D. G.; Ricci, G.; Rossen, K.; Cossy, J. *Org. Lett.* **2013**, *15*, 3876−3879.

CHAPTER 5

Traditional Synthesis of Benzo[b]furans

In this chapter, we provide some good examples of traditional methods of benzofuran synthesis.

In 2009 Tomkinson and coworker reported direct synthesis of benzofurans from *O*-arylhydroxylamines (Scheme 5.1) [1]. This method started with the preparation of *O*-arylhydroxylamine salts. After comparing different synthetic routes, the Sharpless method proved to result in higher yield and was more amenable to scale-up. After condition selection, the reaction used methanesulfonic acid (2 eq.) as the acid and Tetrahydrofuran (THF) as the solvent.

In 2006 Naito reported a new synthetic method for the preparation of benzofurans. The key step of this method is the [3,3]-sigmatropic rearrangement of *N*-trifluoroacetyl-ene-hydroxylamines (Scheme 5.2) [2]. After condition selection, the Trifluoroacetyl Triflate-4-Dimethylaminopyridine (TFAT-DMAP) system was found to be the most effective for constructing various benzofurans.

In 2000, Spoors and coworkers developed a new method for the synthesis of benzofuran (Scheme 5.3) [3]. Firstly, *p*-methoxyphenol was reacted with 2-chloroethylmethanesulfonate and K_2CO_3 in Dimethylformamide (DMF) and gave ether 2 in an unoptimized yield of 50%. Then ether 2 was exposed to two equivalents of bromine in dichloromethane in the presence of iron granules to create dibromide 3 with 81% yield. Finally, the mixture of *n*-butyllithium with dibromide 3 (inverse addition) treated at −40°C resulted in clean conversion to aldehyde 4 with 75% yield after addition of DMF.

In 2012, Wirth and coworker developed a metal-free cyclization of orthohydroxystilbenes to benzofurans (Scheme 5.4) [4]. After condition selection, 1 eq. of $PhI(OAc)_2$ with a reaction time of 2 h in

Transition Metal-Catalyzed Benzofuran Synthesis. DOI: http://dx.doi.org/10.1016/B978-0-12-809377-1.00005-X

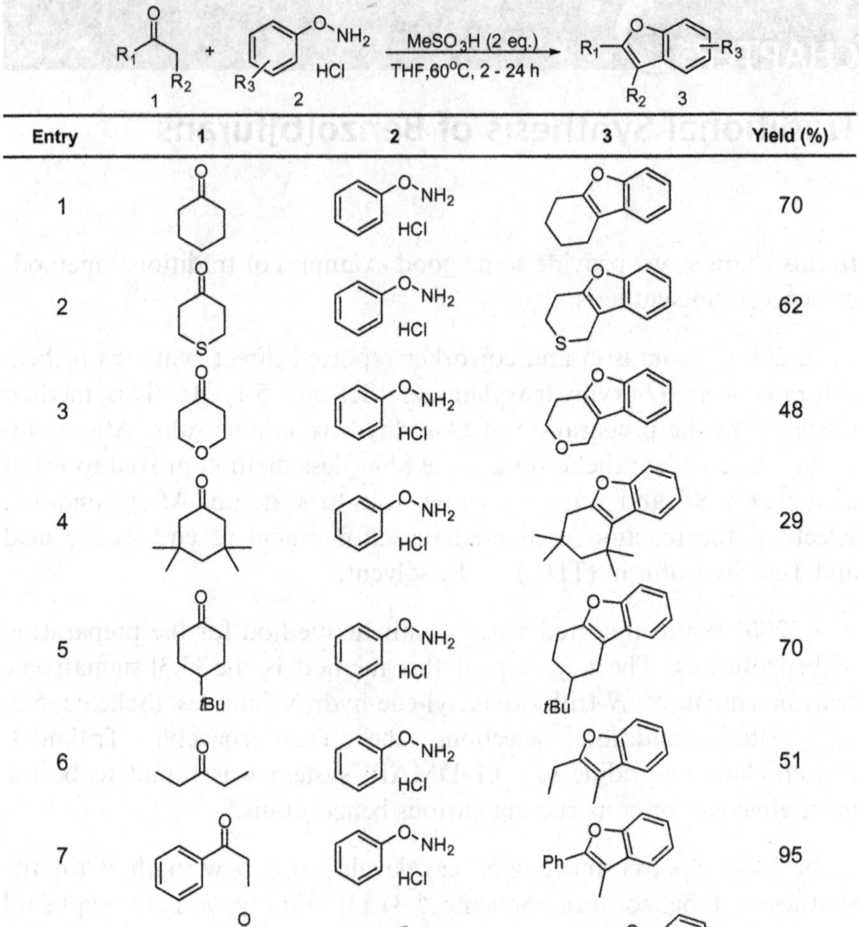
MeSO3H (2 eq.)
THF,60°C, 2 - 24 h
R1
R2
R3
NH2
HCl
1
2
3
Entry
1
2
3
Yield (%)
1 70
2 62
3 48
4 29
5 70
6 51
7 95
8 79
9 84
10 76
tBu
Ph
Br

Scheme 5.1 Synthesis of disubstituedbenzofurans from O-Arylhydroxylamines.

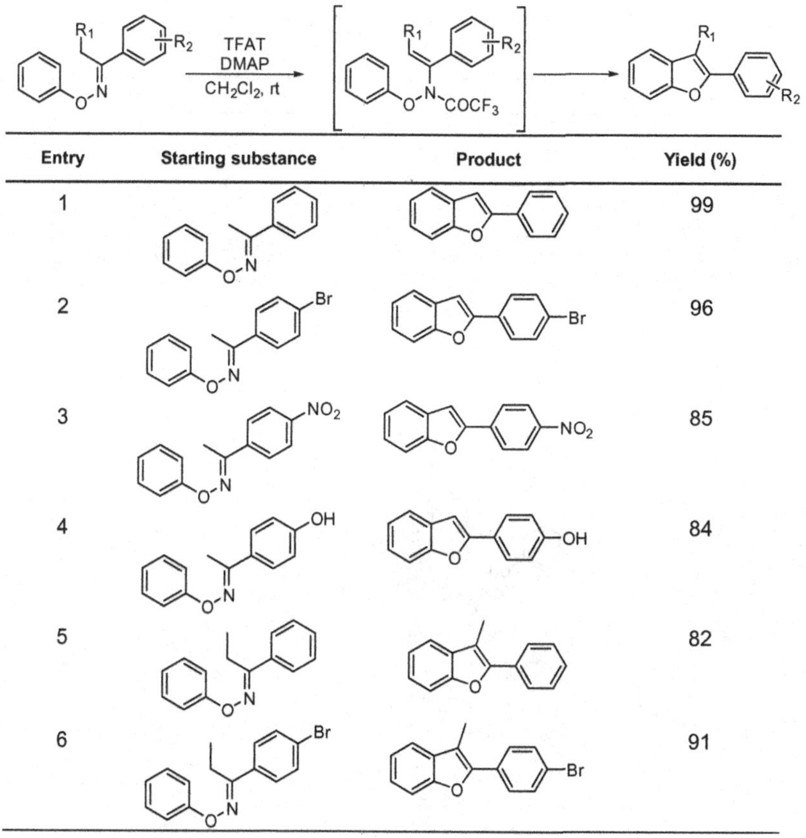

Scheme 5.2 Synthesis of natural 2-arylbenzofurans.

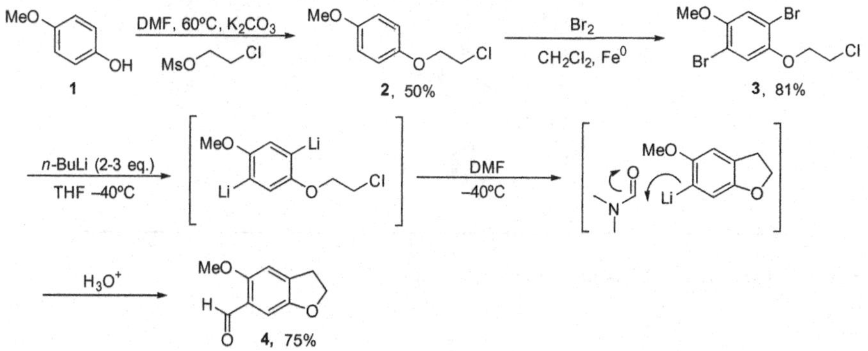

Entry	Starting substance	Product	Yield (%)
1			99
2			96
3			85
4			84
5			82
6			91

Scheme 5.3 Synthesis of benzodihydrofurans.

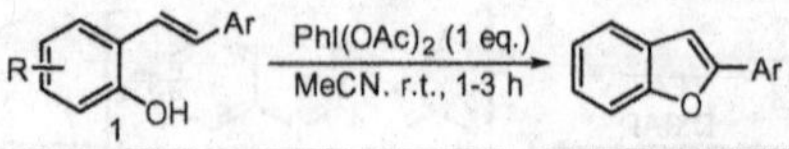

Entry	1	Yield (%)
1	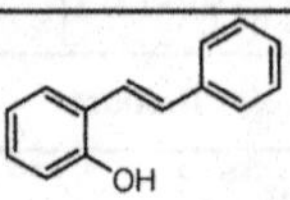	77
2	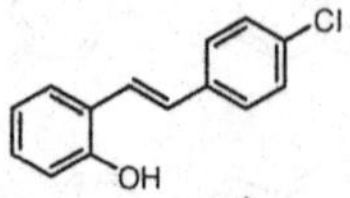	79
3	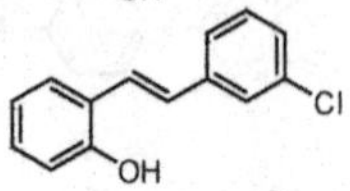	77
4	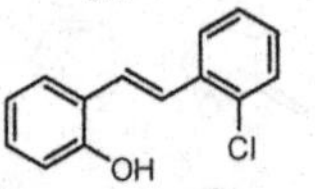	74
5	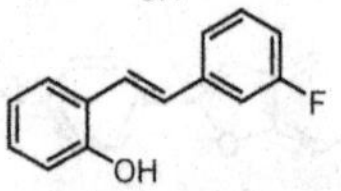	68
6	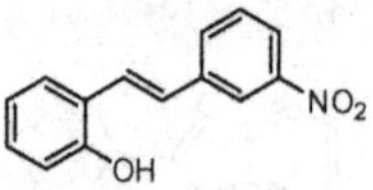	69
7	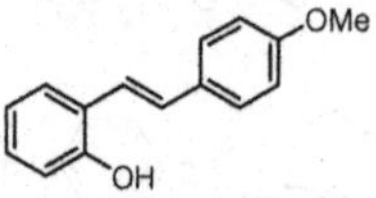	87
8	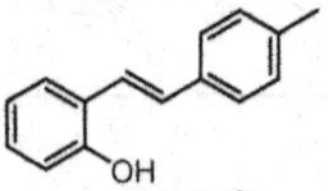	79
9	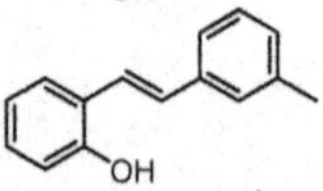	83
10	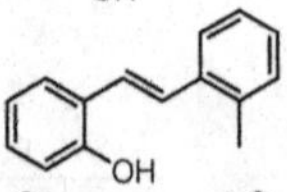	86
11	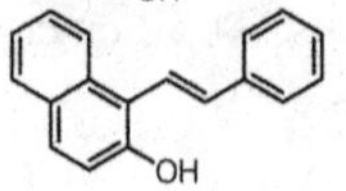	88
12	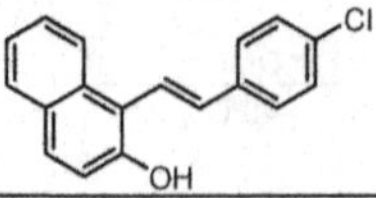	91

Scheme 5.4 Cyclization of (E)-2-hydroxystilbenes.

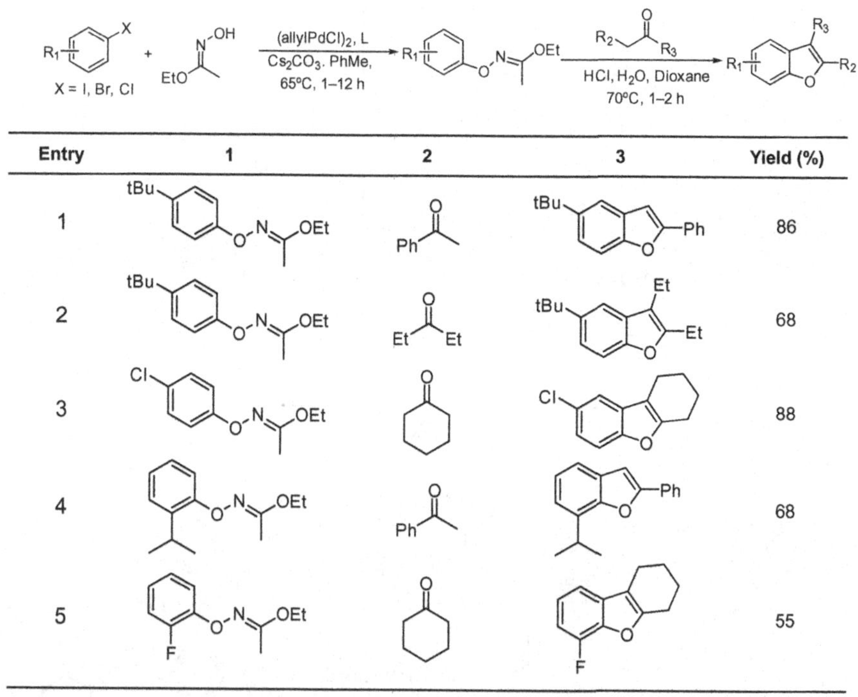

Entry	1	2	3	Yield (%)
1				86
2				68
3				88
4				68
5				55

Scheme 5.5 One-pot synthesis of benzofurans.

CH_3CN were found to be the best conditions for this reaction. This reaction has a wide scope of starting substances. Both electron-donating and electron-withdrawing groups can be tolerated. This is a simple method for the synthesis of benzofuran.

In 2010 Buchwald and coworker developed Pd catalyzed for the O-arylation of ethyl acetohydroximate with aryl chlorides, bromides, and iodides (Scheme 5.5) [5]. These O-arylated products can be used in the synthesis of substituted benzofurans.

In 2007 Giacomelli and coworker reported a facile method for the synthesis of chiral 2-substituted benzofurans (Scheme 5.6) [6]. This method involves two steps. First, 3 eq. of acid 1 or the optically active N-Boc amino acid were treated with 2,4,6-trichloro-1,3,5-triazine (TCT) (1 equiv) and NEt_3 in Dichloromethane (DCM) to form the activated ester 2. The reaction was carried out under microwave irradiation in a sealed tube. Then, toluene, 2-hydroxybenzyl triphenylphosphonium

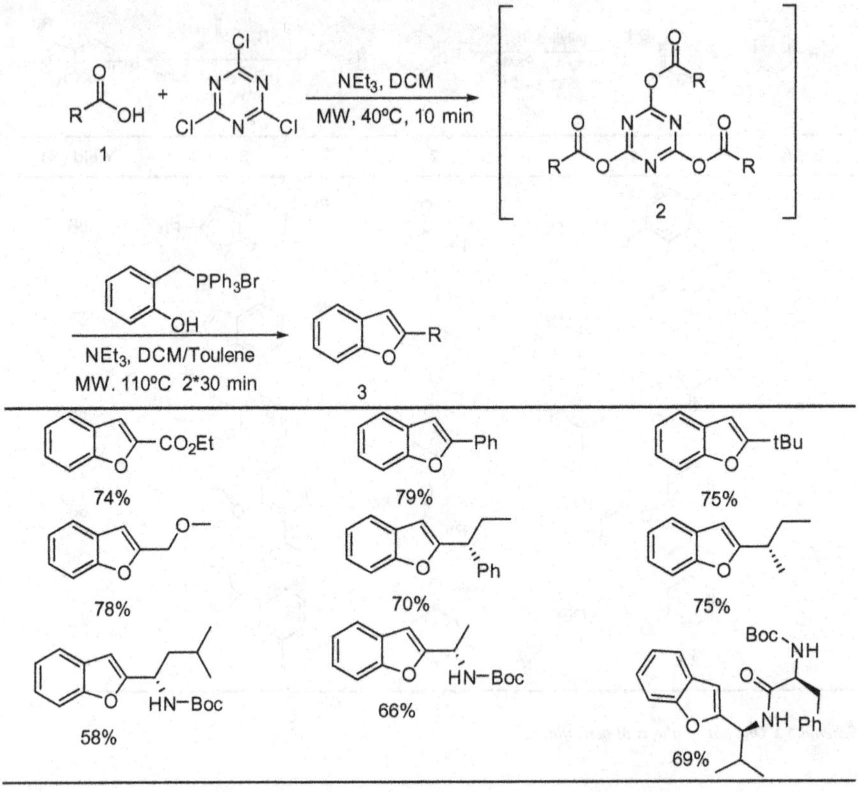

Scheme 5.6 Conversion of carboxylic acids into benzofurans.

bromide (3 eq.), and NEt₃ were added in the cooled reaction mixture and irradiated at 110°C for two cycles of 30 min.

In 2012 Marko and coworker developed a new method for the synthesis of benzofuran by *O*-hydroxyphenones and dichloroethylene involving two steps (Scheme 5.7) [7]. First, *O*-hydroxyphenones react with 1,1-dichloroethylene and generate the corresponding chloromethylene furans. Then, under acidic condition the product of the first step rearranges into benzofuran carbaldehydes. Using this novel method, several benzofurans can be obtained with good or excellent yields.

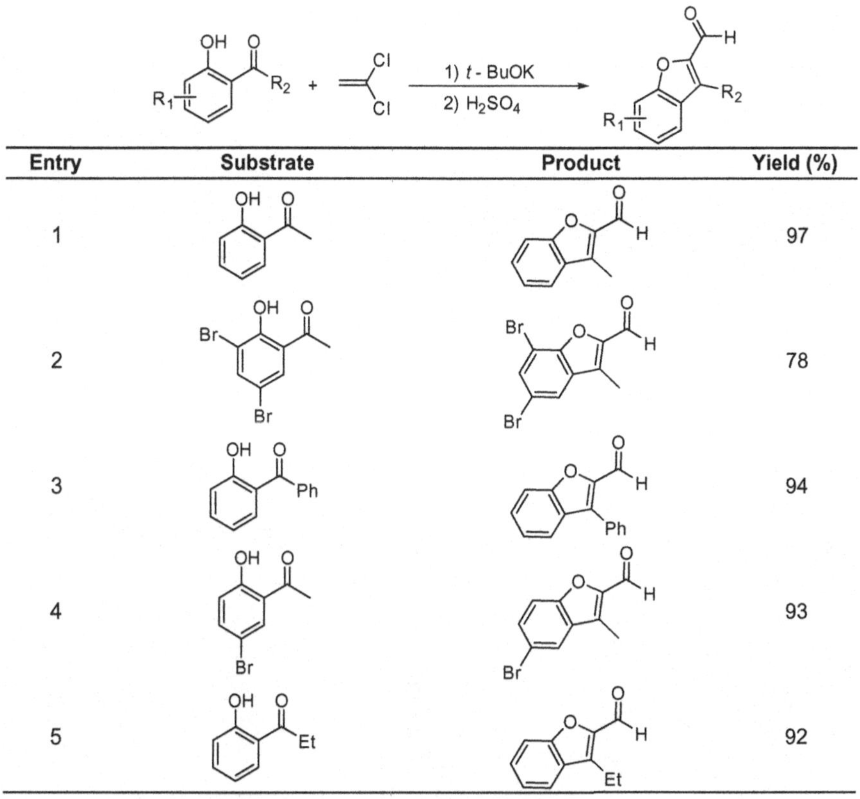
1) *t*-BuOK
2) H_2SO_4
Entry
Substrate
Product
Yield (%)
1
97
2
78
3
94
4
93
5
92

Scheme 5.7 Synthesis of benzofurans.

REFERENCES

[1] Contiero, F.; Jones, K. M.; Matts, E. A.; Porzelle, A.; Tomkinson, N. C. O. *Synlett* **2009,** 3003–3006.

[2] Takeda, N.; Miyata, O.; Naito, T. *Eur. J. Org. Chem.* **2007,** 1491–1509.

[3] Plotkin, M.; Chen, S.; Spoors, P. G. *Tetrahedron Lett.* **2000,** *41*, 2269–2273.

[4] Singh, F. V.; Wirth, T. *Synthesis* **2012,** *44*, 1171–1177.

[5] Maimone, T. J.; Buchwald, S. L. *J. Am. Chem. Soc* **2010,** *132*, 9990–9991.

[6] De Luca, L.; Giacomelli, G.; Nieddu, G. *J. Org. Chem.* **2007,** *72*, 3955–3957.

[7] Schevenels, F.; Markó, I. E. *Org. Lett.* **2012,** *14*, 1298–1301.

CHAPTER 6

Summary and Outlook

The main achievements of benzofuran synthesis have been summarized and discussed here, organized according to the catalyst systems applied. Hopefully this text has been useful to the synthetic community. Please forgive any omissions by the author.

Transition Metal-Catalyzed Benzofuran Synthesis. DOI: http://dx.doi.org/10.1016/B978-0-12-809377-1.00006-1

INDEX

Note: Page numbers followed by "*f*" refer to figures.

Printed in the United States
By Bookmasters